■■ spring into ■■

Technical Writing for Engineers and Scientists

▮▮spring into▮▮ series

Spring Into... a series of short, concise, fast-paced tutorials for professionals transitioning to new technologies.

Find us online at **www.awprofessional.com/springinto/**

Spring Into Windows XP Service Pack 2
Brian Culp
ISBN 0-13-167983-X

Spring Into PHP 5
Steven Holzner
ISBN 0-13-149862-2

Spring Into HTML and CSS
Molly E. Holzschlag
ISBN 0-13-185586-7

Spring Into Technical Writing for Engineers and Scientists
Barry J. Rosenberg
ISBN 0-13-149863-0

Spring Into Linux®
Janet Valade
ISBN 0-13-185354-6

YOUR OPINION IS IMPORTANT TO US!

We would like to hear from you regarding the Spring Into... Series. Please visit **www.awprofessional.com/springintosurvey/** to complete our survey. Survey participants will receive a special offer for sharing their opinions.

From the Series Editor
Barry J. Rosenberg

A few years ago, I found myself in a new job in
which I had to master many new skills in a very
short time. I didn't have to become an instant
expert, but I did have to become instantly compe-
tent.

I went to the bookstore but was shocked by how
much the publishing world had changed. At a
place where wit and intelligence had once been
celebrated, dummies were now venerated. What
happened?

Photograph courtesy of Ed Raduns

Well, I made a few phone calls, got Aunt Barbara to sew up a few costumes, and con-
vinced Uncle Ed to let us use the barn as a stage. Oh wait... that was a different problem. Actually, I made a few phone calls and got some really talented friends to write
books that clever people wouldn't be ashamed to read. We called the series "Spring
Into..." because all the good names were already taken.

With Spring Into..., we feel that we've created the perfect series for busy professionals. However, there's the rub—we can't be sure unless you tell us. Maybe we're hitting
the ball out of the park and straight through the uprights, bending it like Beckham, and
finding nothing but net. On the other hand, maybe we've simply spun a twisted ball of
clichés. Only you can tell us. Therefore, if anything—positive or negative—is on your
mind about these books, please email me at

barry.rosenberg@awl.com

I promise not to add you to any email lists, spam you, or perform immoral acts with your
address.

Sincerely,
Barry

■■ spring into ■■

Technical Writing for Engineers and Scientists

Barry J. Rosenberg

♠Addison-Wesley

Upper Saddle River, NJ • Boston • Indianapolis • San Francisco
New York • Toronto • Montreal • London • Munich • Paris • Madrid
Capetown • Sydney • Tokyo • Singapore • Mexico City

Many of the designations used by manufacturers and sellers to distinguish their products are claimed as trademarks. Where those designations appear in this book, and the publisher was aware of a trademark claim, the designations have been printed with initial capital letters or in all capitals.

The author and publisher have taken care in the preparation of this book, but make no expressed or implied warranty of any kind and assume no responsibility for errors or omissions. No liability is assumed for incidental or consequential damages in connection with or arising out of the use of the information or programs contained herein.

The publisher offers excellent discounts on this book when ordered in quantity for bulk purchases or special sales, which may include electronic versions and/or custom covers and content particular to your business, training goals, marketing focus, and branding interests. For more information, please contact:

U. S. Corporate and Government Sales
(800) 382-3419
corpsales@pearsontechgroup.com

For sales outside the U. S., please contact:

International Sales
international@pearsoned.com

Visit us on the Web: www.awprofessional.com

Library of Congress Cataloging-in-Publication Data

Rosenberg, Barry J.
    Spring into technical writing for engineers and scientists / Barry J. Rosenberg.
        p. cm.
    Includes bibliographical references and index.
    ISBN 0-131-49863-0 (alk. paper)
    1. Technical writing. I. Title.

T11.R663 2005
808'.0666—dc22

2005041173

ISBN 0-131-49863-0
Text printed in the United States on recycled paper Courier in Stoughton, Massachusetts.
First printing, May 2005

To Marilyn, Rachel, and Danny, with all my love

To the memory of Caryl Dundorf:
champion of humanity and tireless role model

Contents

Preface

The character of Fortuna—brilliant, sexy heiress to the Gambini empire, and two-time winner of both the Grand Prix de Monaco and the Fields Prize in mathematics—is strictly fictional. In fact, all the characters in this book—whether living or dead, implied or extrapolated, fictional or nonfictional—are fictional. The people, the companies, the situations, and everything presented as 100% factual are completely fictional. In fact, I don't even really exist; "Barry Rosenberg" is just a composite author formed by the publishing company from several actual technical authors.[1]

To be completely honest, this book contains no character named Fortuna. All that stuff about everything being fictional is fictional. The information in this book is the truth and factual and asymptotically approaches satisfactual. I'm real, too. I'm a real technical writer, manager, and teacher, working in the software industry. I occasionally teach technical writing to engineering and science students at a surreal place called MIT.

Who Should Read This Book?

I've aimed this book at engineers and scientists who must write about stuff.

Perhaps you are an esteemed 60-year-old scientist who has long realized how integral writing is to the job. Perhaps you are a 20-year-old science student who is taking a class in technical writing because "you have to." Perhaps your career is somewhere between those two points, and you find it painful to write the specs and reports that your job requires, and you are sick of your peers scribbling "I don't understand" in the margin of everything you write, and you just wish that there were a way to make writing go a little easier.

Perhaps you are already a good writer and would like to take your writing to another level.

Let me re-emphasize: This book is for engineers and scientists, not professional writers. I've assumed that you don't much care about the difference between transitive and intransitive verbs—you only want to write better. If you'd like to find out more about the demographics for this book and who I think you are, see Table 2-1.

---

1. Much the same way that a nineteenth-century publishing syndicate formed "Mark Twain."

How Is This Book Organized?

I've organized this book into the following four sections:

- Section 1 introduces the field and explains how to plan documentation.
- Section 2 teaches you the nuts and bolts of technical and scientific writing.
- Section 3 explains how to write particular kinds of engineering and scientific documents.
- Section 4 covers editing and producing documentation.

The book concludes with a glossary of writing terms.

Section 1 contains the following chapters:

Section 1—Planning to Write

Chapter	Title	Teaches You How to...
1	The Quest	Understand what this field is all about.
2	Audience	Evaluate whom you are writing for along multiple parameters.
3	Documentation Plans	Create documentation specs and documentation project plans.

Section 2 contains the following chapters:

Section 2—Writing: General Principles

Chapter	Title	Teaches You How to...
4	Words	Choose your words carefully.
5	Sentences	Create accurate, concise sentences,
6	Paragraphs and Sections	Produce proper paragraphs and sections.
7	Lists	Generate professional bulleted and numbered lists.
8	Tables	Create well-organized tables.
9	Graphics	Create technical figures, illustrations, and photographs, and integrate them into a document.
10	Professional Secrets	Master some of the advanced tricks of the writing trade.

Section 3 details specific kinds of documents. It contains the following chapters:

Section 3—Writing: Specific Kinds of Documents

Chapter	Title	Teaches You How to...
11	Manuals	Write five common styles of manuals.
12	Web Sites	Create effective technical Web sites.
13	Proposals	Write research proposals, business plans, and book proposals.
14	Internal Planning Documents	Sell your own company on an idea and produce specs that will lead to good products.
15	Lab Reports	Create lab reports that editors will like.
16	PowerPoint Presentations	Write and deliver a memorable and effective presentation.
17	E-Mail	Write perfectly clear e-mail messages that won't lead to flame wars.

Section 4 contains the following chapters:

Section 4—Editing and Producing Documents

Chapter	Title	Teaches You How to...
18	Editing and the Documentation Process	Put documentation through a proper quality assurance process.
19	Fonts and Typography	Pick the appropriate fonts.
20	Punctuation	Use the right marks in the right places.

What's Unusual about This Book?

This book—like the other books in the Spring Into... Series—provides the following eccentricities:

- Each topic is explained in a discrete one- or two-page unit called a *chunk*.

- Each chunk builds on the previous chunks in that chapter.

- Most chunks contain one or more examples. I believe that good examples provide the foundation for almost all useful technical documents.

- Many chunks contain sidebars and "Quantum Leap" sections, which provide helpful, if sometimes digressive, ancillary material.

I assume that you are a very busy person for whom the time spent in the very act of buying this book was excruciatingly painful. To repay that incalculable opportunity cost, I've adopted the chunk style of presenting information so that you can learn as rapidly as possible.

Finally, I hope you'll find this book fun to read. If you've paid good money for a book—no matter what the topic—boring text is a slap in the face.

Writing a Book about Writing Books

I had this great cognitive psychology professor as an undergraduate. Three times every week, he lectured us on current research on memory. Without fail, in the middle of every lecture, he ran back to his office to fetch the notes he had forgotten. He followed in the same vein as my acne-scarred dermatologist, my cross-eyed opthamologist, and my sister's speech pathology professor, who had a regrettable stuttering problem.

All those people haunted me while I wrote this book. I kept wondering whether I was the writing professor who couldn't write well. After writing each sentence, I stepped back and asked, "Am I practicing what I'm preaching?" Friends, it got ugly. I'd write a sentence, then erase it, then rewrite it, and erase it, and on and on it would go. Writing suddenly became very difficult for me. My self-doubt reached biblical proportions.

Then it hit me—I had become the audience. I had re-experienced the pain of writing. This was a breakthrough because "becoming the audience" is one of the most important states a technical or scientific writer can achieve. Yes, pain is good.

May I write about something else now?

Where Can You Download Examples Used in This Book?

You can download a subset of the examples from this book by browsing to the following URL:

`www.awprofessional.com/title/0131498630`

What Is Fake in the Examples?

I am honor bound to proclaim the following disclaimers about the examples:

- All of the companies mentioned (Dexco Unlimited, Carambola Publishing, Pravda Mills, Googleplex, Calispindex, and so forth) in this book are figments of my imagination. If I accidentally picked the name of a real enterprise, then it was purely a coincidence.

- The sample biographies used in this book are of fictitious people.

- The sample proposals and lab reports exist solely to teach you how to write better proposals and lab reports; they are not based on real proposals or real experiments.

Who Helped Me Write This Book?

Mark Taub—the publisher of this book—wisely appointed the following three primary reviewers, all of whom were completely amazing:

- Mary Lou Nohr—brilliant wit of technical editing—who turned out beautifully detailed and highly humorous responses to my drafts. Mary Lou's comments were, themselves, of publishable quality.

- Chris Sawyer-Laucanno—poet, biographer, expert in ancient languages, and technical writing professor at MIT—who offered insightful and crucial criticism.

- Nicholas Cravetta—engineer and writer—whose tough love kept me on the straight and narrow.

Much of the material in this book originated from a technical writing course I taught for four semesters at MIT. I am indebted to Jim Paradis, Les Perelman, and Steve Strang for giving me the opportunity and the guidance to teach that course.

Julie Nahil did a wonderful job guiding this book through its final editorial phases.

Other material in this book comes from conversations with great technical writers, including Jim Garrison, Marietta Hitzemann, John Abbott, and Judy Tarutz. Special thanks to Kenyon College and to the technical writing department at Rensselaer Polytechnic Institute for preparing me for the technical writing life. Thanks also to Roger Stern and Arthur Lewbel for random props, information, and jokes. Gigantic thanks to the brilliant engineers at 170 Systems, who served as the inspiration for much of this book.

Finally, enormous thanks to my wife Marilyn, who took care of far too many day-to-day details over the last year so that I could have the time to write this book.

Planning to Write

T his section introduces the field of technical and scientific writing, then explains how to evaluate your target audience. It concludes by showing you how to write professional specifications for the documents you plan to write.

CHAPTER 1

The Quest

When I meet someone at a party and say that I'm a technical writer, there is never a follow-up question. After an awkward silence, the person standing next to me invariably becomes the center of attention. (*You shuck oysters for a living? How perfectly droll! Do tell us more.*)

And yet, good technical and scientific communication is one of the building blocks of civilization. Preserving discoveries through writing means that others can benefit from them. Inventing life-saving pharmaceuticals is useless without clear descriptions of how to fabricate and use them properly. How many wonderful discoveries have been lost because a scientist or engineer didn't or couldn't successfully describe the idea? How many commercial ideas have failed because no one could explain how to use the product? How many ideas have failed because readers were bored instead of inspired?

When teaching technical writing to science and engineering students, I've heard them voice concerns such as the following:

- *I like writing fiction, but technical writing is really uncreative and dull.* Indeed, technical writing can be dull, but it doesn't have to be.

- *Engineers and scientists can't write.* This is a "truism" that just isn't true. In fact, many technical people write beautifully.

- *Writing is really hard.* Okay, this one is true.

Technical Writing Theorems

If you asked a professional technical writer to recite the key principles of technical writing while standing on one foot, she would probably say something like the following:

- Write appropriately for your audience.

- Write clearly.

- Write concisely.

- Engage your audience.

- Help the reader.

These theorems often conflict with each other. How, for example, can you describe something accurately while staying as concise as possible? How do you engage your target audience while describing something as dry as dust? To solve these conflicts, you'll need to enhance a seemingly conflicting pair of skills—creativity and discipline.

Technical writing offers few theorems. These theorems lead to only a few dozen corollaries. For example, the "write concisely" theorem leads to corollaries such as the following:

- Keep sentences short by eliminating unnecessary words.

- Write sentences primarily in active voice.

- Eliminate unnecessary sentences or irrelevant concepts.

Since technical writing offers so few fundamental theorems and corollaries, it would seem to be an easy art to master. Although, if you were to apply this same logic to painting, you could become an expert painter simply by mastering three colors and a few kinds of paint brushes. As with painting, though, correct application of the simple theorems takes years of guided practice. This book offers a good start, but it is only by practicing with an experienced editor that you will truly learn to write well.

The following ditty—a paraphrase of a famous quote on diplomacy by Isaac Goldberg—summarizes your quest:

> Technical communication is to write and to say
> The geekiest things in the simplest way.

Technical Writing Can Be Creative

Technical and scientific writing has a well-founded reputation for being dull. Consider the following passage, which is aimed at lay readers:

> Hot air is less dense than cold air. When you put hot air on a scale, it weighs less than cold air. Consequently, hot air rises and cold air sinks. Since hot air and cold air frequently come into contact with each other in the atmosphere, a lot of mixing of air results, which can cause interesting reactions.

The preceding passage meets a few of the criteria for good writing. It is fairly appropriate for the target audience (though lay readers might have trouble imagining the weight of air pressing against a scale), and it is concise. However, the biggest sin in the preceding passage is that it does not engage the audience. It produces no spark; you can almost hear the reader snoring.

Despite its reputation, technical and scientific writing can be highly creative and engaging. It is really a matter of mind-set—you must truly feel the obligation to go beyond a dry recitation of facts to provide a treat for your audience. Injecting life into the preceding passage yields the following:

> Did you ever watch a hawk fly over a highway on a hot day? He flies so easily, barely beating his wings at all. That's because the air over the highway is very hot. Since hot air rises, the hawk gets a free ride up. Conversely, cool air sinks. So, when the hawk flies over cooler terrain, he has to beat his wings extra hard.
>
> On a larger scale, when a big mass of hot air meets a big mass of cold air, our hawk must take shelter from stormy weather.

The second passage is far more engaging than the first for the following reasons:

- Air is too abstract to visualize, but a hawk bobbing up and down on the air is easy to visualize.

- Many readers find stories about animals compelling.

- The image of the soaring hawk provides a mnemonic device for this lesson.

For all forms of communication, remember this lesson:

> You cannot impart information unless you have the audience's attention.

Tell 'Em

I saw a famous folksinger in concert a few years ago. Although he has had a wide-ranging career spanning many decades and musical styles, his name is instantly identified with a single song that he wrote long ago. In concert, he sang a few songs and then mentioned that—before he could continue—he had "a little piece of business" to attend to. Then, with his face twitching, he scratched out a limp rendition of his most popular song.

Before continuing, there is a little piece of business that I must attend to. It goes something like this:

1. Tell 'em what you're going to tell 'em.

2. Tell 'em.

3. Tell 'em what you told 'em.

The preceding ditty—a folksy version of "introduction, body, conclusion"—is undoubtedly technical writing's greatest hit. Many engineers and scientists recite this formula with an "$F = ma$" certainty. It is a useful formula for organizing writing at the following levels:

- At the book level, the opening chapter should introduce the book and the closing chapter should summarize the book's contents.

- At the chapter level, the opening section should introduce the chapter's topic and the closing section should summarize the chapter.

Some writers use this formula right down to the paragraph level. Indeed, in certain kinds of writing, paragraphs do benefit from a classic strong opening and closing sentence. However, good technical prose emphasizes speed, so the formula imposes too much baggage to be practical at the paragraph level.

Experimentation shows that audiences tend to remember facts presented at the beginning and the end of a session better than in the middle. Thus, it is a sound idea to highlight key principles in both the introduction and conclusion. Furthermore, repetition is essential for learning.

The downsides to this formula are as follows:

- Repetition is not concise.

- Many writers produce wasteful conclusions that are almost identical to the introduction.

- Many readers now skip over summary sections (except in lab reports).

The Value of Technical Communication to You

The difference between success and failure in the technical world is generally the ability to communicate effectively.

In engineering and science, *brilliant* does not always equate to *successful*. In fact, *brilliant* often equals *frustrated*. Of course, success is a very tricky concept to define—one engineer's success might equal another's failure. Nevertheless, I'll define engineering success as having one or more of the following parameters:

- The engineer is promoted to a management position.

- The engineer is paid significantly more than his or her peers.

- The engineer invents something that changes the world in a positive way.

- The engineer generally gets his or her way in technical disagreements. (Some engineers find this parameter more precious than oxygen.)

In my opinion, successful engineers convey their ideas better than unsuccessful engineers. Top management positions in engineering organizations are staffed by those who communicate more effectively and more persuasively than their brilliant colleagues. Due to the rising use of e-mail and instant messaging in technical organizations, writing clearly and concisely has never been more important.

Comparing Technical Writing to Engineering and Science

Technical writing shares the following *similarities* with engineering and science:

- **Successful projects require careful planning.** Good engineering teams produce detailed engineering specifications prior to implementation. Good technical writers produce detailed documentation specifications prior to writing.

- **Successful projects require testing.** Good engineering projects run beta tests to find defects in early versions of the product. Good technical writers beta test early versions of documentation for feedback.

- **Successful projects require iteration.** No matter how much planning precedes implementation, engineering typically requires a certain amount of iteration to achieve perfection. An old documentation adage recursively states, "Writing is rewriting." Good writers and good engineers constantly ask, how can I improve this?

- **The principle of parsimony is paramount.** In both engineering and technical writing, simpler is better.

- **Achieving parsimony is difficult.** Elegance is not cheap. For example, in software engineering, it usually takes much longer to code an algorithm in 100 lines than in 200 lines. Similarly, it can take a lot longer to document a complex idea in 100 pages than in 200 pages.

Technical writing *differs* from science in the following ways:

- **Writing cannot be successfully reduced to formulas.** The goal in most sciences is to create mathematical equations that model and predict phenomena. In technical writing, although readability formulas can estimate a passage's audience level, no formula can predict whether writing is clear. Ultimately, the only true test of a document is getting humans to read and act on it.

- **Writing and science require different thought processes.** Scientists construct paradigms from variables and constants. Writers construct documents from words. Both pursuits are logical, but the two disciplines use different parts of the brain. Scientists seek objective logic; writers seek a subjective empathy with their readers.

CHAPTER 2

Audience

Realtors have an old joke that goes something like this:

> Q: What are the three most important things in real estate?
>
> A: Location, location, and the third one isn't really that important.

The technical writing variant on the joke is as follows:

> Q: What are the three most important things in technical writing?
>
> A: Audience.

Get it? See, technical writers are really concise, so... ahh, forget it.[1] At any rate, identifying your audience's needs is essential for writing good documents. This chapter helps you define your audience through a set of questions, such as the following:

- What does my audience already know about this technology?

- What is the native language of my audience?

If you work for a company, your marketing department should be able to help define your audience. Large companies often have extensive market-research data.

Picturing Your Audience

The concept of audience is too abstract for some writers. To solve this problem, consider pasting a picture of someone in your target audience up on your monitor. Some writers find it easier to write for an individual than for a collection of people.

---

1. The realtors' version is funnier. I weep for my profession.

General Education Level

How much general education has your audience had? Did your audience attend high school? College? Graduate school?

Readers prefer text that is appropriate for their general education level. Writing too "high" makes readers feel dumb. Writing too "low" makes readers feel that they are wasting their time.

Most textbooks on writing tell you to write long sentences for highly educated readers and short sentences for poorly educated readers. However, lengthy sentences have no place in technical prose; the goal in technical writing is always to produce short, clear sentences.

Most textbooks on writing also tell you to use vocabulary appropriate for your readers' educational level. In other words, use a broad, sophisticated vocabulary for highly educated readers and a limited, unsophisticated vocabulary for poorly educated readers. After all, you don't want uneducated readers scrambling for their dictionaries after every sentence.

Many fiction authors often ignore this principle, preferring beautifully ornate words like *labyrinth* and *absinthe* to compose sentences such as the following:

> The ornately beautiful carriage bumbled chaotically through the morass like a butterfly drunk on absinthe careening through a labyrinth.

The preceding sentence is, indeed, ornately beautiful; however, figuring out what it means (or whether it means anything at all) takes a while. By contrast, good technical prose is built for speed; technical readers do not have the patience to parse exceedingly complex sentences. Nevertheless, if you are comfortable, you can certainly mix in bona-fide "big" words when your audience is well educated. Sometimes a big word is the most concise word since it may spare you from using multiple little words in its place. On the other hand, if your writing vocabulary is somewhat limited, you should not force yourself to use high-falutin words just because your audience is highly educated. In technical writing, the perfect word is often the simplest word.

Experience and Expertise

How much experience does your audience have with this technology or topic? Does your audience already have formal training in this technology?

If your audience is already familiar with the topic, you can start at a higher level than if they are new to it. Never disrespect a trained audience by explaining topics that they have already mastered. When writing for an experienced audience, you can use jargon and acronyms freely. Conversely, if your audience lacks experience, you must carefully define all technical terms. For example, when writing for an audience of professional programmers, the following passage is appropriate:

The product provides an extensive Java API.

Providing the same information to nonprogrammers would yield a passage such as the following:

The product provides an extensive Application Programming Interface (API). Programmers can use this API to extend the features of our product or to integrate other software products with our product. The APIs for our product are written in Java, which is a popular programming language.

Alternatively, many nonprogrammers do not care about technical details, so a passage like the following (which completely omits the acronym *API*) might be just as effective:

We provide manuals describing how your programmers can extend the product's features. In other words, if our off-the-shelf product isn't exactly what you want, your programmers can customize it.

Sometimes your audience has had plenty of negative experience. Is your audience skeptical of this technology? Never ignore negative expectations. For example, consider the following inoculation from a cell phone manual:

You may have run into problems getting a signal from your previous cell phone. In some cases, this was due to a lack of cell towers. Our continentwide network of cell towers ensures that you are always in range. In other cases, your old cell phone may have lacked the power to transmit a strong enough signal. Thanks to improvements in battery technology, your new cell phone transmits a much stronger signal.

Breadth of Audience

Are you writing for a tightly focused audience or for a broad range of readers?

In some situations, you can pin your audience down to a fairly precise demographic. For example, when you are writing a lab report on an esoteric topic, all your readers are almost certainly experts on the topic.

In other situations, your audience is rather broad. For example, consider the breadth of audience for a cell phone user's manual. The readers span a wide age and educational background. Some readers have never used a cell phone before, and others have used them daily for 15 years.

In still other situations, the audience's breadth falls somewhere in between. For example, consider a manual explaining how to administer a Windows system. The core audience for this manual will probably be professional system administrators with a fair amount of expertise in Windows. However, many amateurs might also use this manual to learn how to administer their PCs at home.

It is far harder to write for a broad audience than for a narrow one. When writing for a broad audience, you should do the following:

- Aim the bulk of the documentation at the least experienced readers.
- Aim a few advanced chapters or sections at experienced readers. Label these chapters or sections explicitly as "For Advanced Readers."

If time and money are no object, your organization should handle a broad audience by writing two separate documents:

- one aimed at inexperienced readers
- one aimed at experienced readers

Native Language

What percentage of your audience reads English as a native language? How well does your audience actually read English?

Writing for Technologists

English has become the dominant technical and scientific language in the world. Most technologists and scientists worldwide can read English. In many European countries, technical audiences comprehend technical prose written in English almost as well as audiences in the United Kingdom or United States.

How Pervasive Is English?

Two colleagues—a Danish engineer and a Swedish engineer—hold a weekly teleconference. (Danish and Swedish are similar languages.) I asked the Danish engineer which language was spoken during the teleconference. He replied, "I speak to him in Danish. He answers me in Swedish. And when we are having trouble understanding each other, we speak English."

Many languages—Chinese, for example—are markedly different from English. In such countries, technologists often have more difficulty understanding English text than in European countries. Many nonnative readers will miss nuances. Note that small misunderstandings about technical text can lead to disastrous technical problems.

When writing for a technical audience containing many nonnative speakers, follow these guidelines:

- **Follow exact grammatical rules—even the annoying ones.** For example, when writing for native speakers, it is okay to split infinitives. However, split infinitives can confuse some nonnative speakers. For example, consider the following passages, which are similar but not quite identical. Version 1 contains a split infinitive but version 2 does not. Nonnative speakers will generally find version 2 far easier to understand than version 1.

1. The filter causes high-frequency notes to gently and sweetly pass.

2. The filter causes high-frequency notes to pass gently and sweetly.

- **Avoid uncommon acronyms.** (It is fine to use common acronyms.) Native English speakers can often guess the meaning of uncommon acronyms, but such acronyms cause problems for nonnative speakers.

- **Simplify vocabulary.** Generally speaking, nonnative speakers have a somewhat smaller English vocabulary than native speakers.

What Is a Simple Word?

Which of the following verbs is simplest?

- use

- utilize

- employ

Native English speakers will certainly pick "use" as the simplest. It is, after all, the shortest and the most commonly... er... used. Native French speakers, however, might choose differently because the verbs "utilize" and "employ" are cognates for common French verbs.

Writing for Nontechnical Nonnative Speakers

When writing for a nonnative audience of nontechnologists, you cannot assume that your readers understand English. True, many of your readers (particularly Europeans) will understand English very well, but many readers will not. In such situations, your organization must translate the documentation into your readers' native languages.

For example, the documentation set for a large software product might consist of the following:

- Manuals for programmers and system administrators, which do not have to be translated.

- Manuals and online help for end users, which do have to be translated.

In certain situations, documentation can consist entirely of graphics, which would not have to be translated. For example, directions for unpacking a shipping container might be given through pictures only.

Native Culture

What percentage of your audience will understand your cultural references?

Writers sometimes mistakenly assume that because readers share the same language, they also share the same culture. The key rules for producing useful technical or scientific prose for a worldwide audience are as follows:

- Avoid slang.

- Pick examples and metaphors that will make sense in multiple cultures.

- Be sensitive to fundamental differences in presentation.

- Don't offend.

Slang

Even among native English speakers, slang is often confusing—Brits, Americans, and Australians do not always understand each other's lingo. (What in the world are *bangers and mash* anyway?)

Baseball is a huge part of American slang. A sentence such as the following makes perfect sense in baseball-loving countries such as the United States and Japan:

Project Walrus hit a home run for the company.

British readers, of course, would say that a baseball metaphor just isn't cricket.

Many writers feel that soccer/football terms are cross-cultural since the game is "universal." Be aware, however, that many American readers would rather eat a soccer ball than go to a soccer game and that soccer slang is not okay for an American audience.

Okay Is Okay

Mysteriously, just about everyone seems to know the meaning of the American slang word *okay* or *OK*. In fact, this word is so okay that it has crept into the slang of other languages.

Examples

Despite assurances to the contrary, it turns out that the world is actually an awfully big place. It is hard to create examples that will make sense everywhere in the world. For example, consider the following seemingly benign passage:

> When an electric force is applied, the charge in a capacitor increases, much the way the balance in your checking account increases when you work. Then, when the charge is needed, the capacitor releases it, much the way you release the value in your checking account by writing a check.

Is the preceding example cross-culturally correct? Unfortunately it is not. Personal checking accounts are unknown in Japan. Before investing too much energy in an example, do some research.

An Almost-Universal Cultural Reference

Earthlings do not share a common religion, language, or political system. However, we do share a knowledge of American movies. Many readers are unfamiliar with rock gardens in Kyoto or tea plantations in India, but most know who R2D2 is.

Presentation

We often forget that many daily details are not universal. For example, consider the following date:

1/3/05

American readers interpret the preceding date as January 3, 2005. European readers interpret it as 1 March, 2005. To prevent cross-cultural confusion on dates, it is clearer to spell out the month as in the following example:

1 March 2005

Some countries present monetary values as integers only, while other countries express money with real numbers. Some cultures use commas to separate groups of three digits in a large number, but other cultures use periods. To avoid confusion, you may need to provide multiple examples of the same value.

Don't Offend

Cross-cultural prose tries not to offend. For example, humans in photographs should be modestly dressed, and any food should be vegetarian. Obviously though, you must be practical. For instance, when documenting high-fat diets, it is fine to depict meat.

Audience Motivation

Audiences read fiction because they want to and technical prose because they have to.

Writers often forget to ask themselves why readers are picking up a document in the first place. The following list provides a few general answers:

- People read manuals to learn how to do something. Readers study manuals to learn how to see through a new telescope, cook with a new microwave oven, or write code in a new programming language.

- People read reference manuals to find information very quickly. Audiences often read reference material to recall a fact they already know.

- Scientists read lab reports to keep up with their field, to evaluate colleagues' progress, and to consider their next experimental steps.

When planning a document, always consider your readers' motivations.

What Is Your Audience's Emotional State?

Quickly now—what is the emotional state of someone reading the "How to Change a Flat Tire" section of an automobile owner's manual? How maniacal is the reader who is looking up fatal error codes in the back of a software manual? How well will a patient who has just received a distressing diagnosis be able to comprehend a doctor's instructions?

Entire manuals will never be read until disaster strikes. When writing about troubling situations, remember the following:

- **Agitated readers are not always thinking rationally.** You might need to start certain sections by reassuring the reader that the current problem is fixable.

- **Agitated readers are probably in a highly impatient mood.** You must provide concise answers.

To help agitated readers, consider the following suggestions:

- Provide a "Troubleshooting" section in manuals.

- Make sure that various kinds of emergencies have entries in the index and table of contents.

Medium and the Message

Is your target audience reading the document in hard copy or soft copy? If in soft copy, does the technology permit hyperlinks? Will people read this document linearly or in a random-access fashion?

I don't honestly believe that the medium is the message. (I lean towards the message being the message.) Nevertheless, the wise writer is aware of the medium through which the document will be read.

One obvious advantage of writing for the Web is that you can provide hyperlinks. With hyperlinks, you can rely on other sources to tell parts of the story. Hyperlinks are particularly useful when your audience is broad. They can redirect newcomers to definitions or introductions without destroying the focus of your prose.

If you are writing for the Web, how will readers arrive at your document? Will they start with your document, or will a hyperlink carry them there? Restating the question, how much will your readers already have read about the topic prior to hitting your document?

Hard copy isn't dead—not yet, anyway. Although paper offers many disadvantages, it does offer much higher resolution than soft copy (up to two orders of magnitude, depending on how you count). Thus, most images will look better on paper than on a Web page. Many readers find long documents much easier to read in hard copy than in soft copy. Also, because of its relative permanence and tangibility, hard copy strikes many readers as being more solid and reliable. Legal transactions generally still require paper. Paper just plain strikes people as being more "real" than soft copy.

Is your audience going to read your document straight through, from cover to cover? Probably not. Modern audiences just do not have the time or patience to read documentation straight through. Most technical readers snack, nibbling on a page here and a chapter there. That is not to say that straight, linear access is dead. Fiction, after all, is still going strong. However, thanks to the World Wide Web, most technical readers prefer random access that takes them straight to the desired fact.

Depending on the medium, you might have the freedom to use attention grabbers such as quotes, sidebars, or photographs to pull in readers.

Becoming the Audience

In teaching a task that requires great concentration, teachers sometimes tell students to imagine themselves in the role of the object being manipulated. Thus, the sculptor imagines the sculpture as an extension of herself, and the archer "becomes" the arrow. In technical writing, the wise author periodically becomes the audience. In other words, the author periodically stops writing and starts reading and relating to the prose as the audience would.

Gaining empathy for the reader is probably the hardest skill for a writer to learn. You must simultaneously be of two minds—the expert and the novice. For example, while documenting how to operate a new medical instrument that you invented, you must imagine what it is like to be the doctor who has never seen this device before. You must frequently step back and ask yourself, if I were in this doctor's shoes, would I understand this?

The best way to become the audience is, literally, to become the audience. That is, you must do the things that your audience would do. For example, suppose a software engineer creates a new Application Programming Interface (API). The classic writer's mistake would be to document only how this API works instead of documenting how another programmer would make practical use of it. To document the API, the writer should first become the audience by using the API to code sample real-world applications. In other words, the writer should use the API exactly as her readers would. Through this experience, the writer would learn not only how the product should work, but she would also learn various limitations on using this product.

Walking a Mile in a Reader's Shoes

It is sometimes impractical to become the audience in a literal way. Nevertheless, you can still gain empathy by reading similar documents immediately prior to writing your own document. For example, suppose you must write a spec for your colleagues. Immediately prior to writing, consider reading a couple of old specs that your colleagues wrote. What did you like about their specs? What confused you? What irritated you? When you write your design spec, emulate the good parts and eliminate the bad. Feel the pain so that your readers don't have to.

Becoming truly empathetic is difficult. After all, it is a long round-trip from expert to novice and back again. This is one of the primary reasons that writing is often best done by someone other than the inventor.

Summary of Audience

As you prepare to write, it is helpful to fill out an audience chart such as that shown in Table 2-1. I've taken the liberty of filling it out based on the planning I did for writing this book. Note that I had the advantage of extensive market research on the target audience.

TABLE 2-1 Target Audience for This Book

Question	Answer
How much general education has the target audience had?	The target audience on average holds a master's degree, mainly in a technical or scientific discipline.
How much experience does the target audience have with technical or scientific writing?	Everyone in the target audience has had at least some technical writing experience, either at school or at work. A few readers have had a lot of experience. Almost no one in the target audience is an expert.
What is the target audience's predisposition toward the subject matter?	The majority of readers find technical and scientific writing to be a chore; a smallish minority enjoy it.
Is the target audience tightly focused or broad?	The target audience is reasonably focused in terms of mathematical and analytical aptitude.
What percentage of the target audience reads English as a native language?	Approximately 85% reads English as a native language. The remaining 15% are technical people who are comfortable reading (and writing) English.
What percentage of the target audience will understand my American cultural references?	Approximately 75% of the target audience lives or has lived in North America. Since 25% live elsewhere, I must be careful about cultural references.
What kinds of examples will the target audience prefer?	The target audience prefers scientific examples. However, since the audience comes from a wide variety of disciplines, I should use examples about general scientific principles that most readers will know.
Why is the target audience reading this book?	The target audience wants to enhance its writing skills, perhaps to take care of a specific short-term assignment or perhaps to achieve longer-term career goals.

 CHAPTER 3

Documentation Plans

Most professional technical writers produce the following two types of documentation plans:

- documentation specifications (doc specs), which detail a single document
- documentation project plans, which detail a set of documents

Documentation professionals ostensibly write these plans to provide a formal feedback mechanism for other people in an enterprise. In other words, these plans provide a way for the head of engineering or a marketing representative to sway what will be written before it is written. Indeed, some organizations really do review documentation plans very carefully and integrate their review into the broader engineering development cycle. However, my experience suggests that you should write documentation plans for a more valuable reason:

Documentation plans lead to better final documents.

Even if no one reads them (hey, it could happen), writing detailed documentation plans will still help you, the writer.

This chapter teaches you how to write both kinds of planning documents.

Document Specifications (Doc Specs)

Before writing a document longer than 25 pages or so, you should create a **document specification (doc spec)**. A good doc spec serves the following three purposes:

- It helps you organize your thoughts about the document.

- It communicates your intentions to the other members of the engineering team.

- It helps your team reach consensus on the purpose and scope of the document *before* you start writing. The engineering adage about measuring twice and cutting once also applies to documentation.

Ideally, creating a doc spec involves the following three-step process:

1. You create a preliminary doc spec.

2. Your reviewers analyze the preliminary doc spec.

3. You generate a final doc spec based on your reviewers' comments. If your reviewers disagree with each other, hold a meeting to iron out differences.

Doc specs need not be long. In fact, short doc specs generally yield a better harvest of comments than long ones. A doc spec for a 50-page manual should only be about two or three pages long. A doc spec for a longer manual should be only slightly longer (due to lengthier outlines). Each doc spec should contain the following sections:

- a brief overview of the project

- a detailed description of the target audience

- a brief description of the nongoals—topics that *won't* be covered in the document

- a section entitled "What Readers Will Know after Reading This Document

- an estimate of the length of the final document

- the tools you will use to write the document, plus the document's output media (PDF, HTML, and/or hard copy)

- optionally, a brief discussion of some of the finer points, such as tone and pace

- a fairly detailed outline, preferably down to first-level heads within each chapter

- a list of reviewers and their responsibilities

- outstanding issues

- a schedule (omitted from the sample doc spec that follows)

Doc Specs: Sample

Our engineering team plans to roll out the Carambola 3000 Weather Station in May. This doc spec describes a manual entitled *Installing the Carambola 3000 Weather Station.*

Target Audience
We are aiming the 3000 at the high-end amateur market. This market is split as follows:

- About 80% of this market consists of hobbyists who are serious enough to spend about $200 on a high-quality weather station.

- About 20% of the audience consist of community organizations such as schools that need a weather station but do not have the budget for a professional model.

The hobbyists know what they are doing mechanically; most of them will look at the schematic diagram and skip the text. The community organizations will require the detailed text.

The reader must also install and configure the software. We assume that all readers are comfortable installing software from a CD onto Microsoft Windows; however, we do not assume any knowledge of computers beyond that. So, the guide must patiently explain configuration.

Nongoals
Our target audience already understands meteorological concepts, so the book will make no attempt to explain weather terms or weather forecasting.

What Readers Will Know after Reading the Manual
After reading this manual, readers will know how to do the following:

- Install the tangible parts of the weather station.

- Install and configure the accompanying software.

- Troubleshoot a malfunctioning weather station.

Length
The book will consume approximately 40 pages, including the front and back matter.

Tools and Output Media
We will write the book with Adobe FrameMaker 7.1 and produce diagrams with Microsoft Visio Professional 2000. After finishing, we will generate a PDF, which we will FTP to our

print vendor. Our print vendor will produce an initial run of 3,000 copies on a 6.5 × 9 trim with a wire-o binding. We will kit the manuals with the product on the assembly floor.

Outline

The following is the preliminary outline (subject to change):

- Front matter (title page, copyright page, table of contents, preface)
- Chapter 1: Overview
    - Capabilities
    - List of Parts
    - Schematic Diagram
- Chapter 2: The Weather Station
    - Control Box and Power
    - Anemometer Unit
    - Hygrometer/Thermometer/Barometer Unit
    - Rain Gauge
- Chapter 3: Software
    - Installation
    - Basic Configuration
    - Connecting to the Internet
- Appending A: Troubleshooting
- Back matter (index, tear-out warranty card)

Reviewers

The following people have been graciously volunteered to review the manual:

- Eswar will review Chapter 1 and 2.
- Andy will review Chapter 3 and Appendix A.
- Martina will perform a literary edit.

Outstanding Issues

We cannot write Chapter 3 until the software has been developed. Given the software development schedule, we will have limited time to write Chapter 3.

Marketing must finalize the new tear-out warranty card and terms prior to October 15.

Documentation Project Plans

A doc spec details a single document; a documentation project plan summarizes an entire documentation set, explaining how different pieces of the puzzle fit together. If your team is writing multiple documents, you should create a documentation project plan. Like a doc spec, a good documentation project plan communicates your intentions to the rest of the team and helps bring the team to consensus.

One way to plan a documentation set is to identify all the audiences you must satisfy and then to figure out which title(s) they need. In the best possible world, each type of user would find everything in a single title. However, this is not always possible. For example, the same audience often needs different kinds of information at different stages, which necessitates multiple titles.

Another way to plan a documentation set is to use a spreadsheet as follows:

1. Create a comprehensive list of *everything* that users need to do, one item per row.

2. In a separate column, identify the audience for each row.

3. Sort the rows by audience.

4. Group all the items related by audience into a title, or possibly, into multiple titles.

> ### How Many Documents Are Best?
>
> Many companies create more titles than necessary. Smart companies minimize the number of titles. You might well ask, why would creating five 200-page books be better than creating ten 100-page books? Producing and maintaining a title (any title, regardless of length) generates various fixed costs, so reducing the number of titles will reduce the total cost.

A documentation project plan needs to express the following:

- the titles in the documentation set, preferably arranged by target audience

- a few sentences about each title

- a media plan, explaining how each title will be delivered to the target audience

- any outstanding issues

- a schedule (omitted from the sample documentation project plan that follows)

Documentation Project Plan: Sample

Our engineering team plans to roll out the Carambola 3000 Weather Station in May. This documentation project plan summarizes the planned documentation set for the product.

The Documentation Set

The documentation set will consist of the titles listed in Table 3-1.

TABLE 3-1 The Titles in the Carambola 3000 Weather Station Documentation Set

Audience	Full Title	Title Abbreviation
Customers	Planning for the Carambola 3000 Weather Station	Planning
	Installing the Carambola 3000 Weather Station	Installing
	Release Notes	Release Notes
Integrators	Developing Applications for the Carambola 3000 Weather Station	Developing
Customer Support Reps	Advanced Troubleshooting for the Carambola 3000 Weather Station	Advanced Troubleshooting
Manufacturing	Schematics	Schematics
	Hardware Specifications	Specs

We detail each of the preceding titles in separate doc specs, which you can find on the corporate intranet.

For Consumers

We will provide the following three titles for customers:

- *Planning*—This guide will help consumers prepare for an installation during the three-week interim between ordering the station and receiving it. This guide will help consumers determine the most effective locations for instruments.

- *Installing*—This guide will help consumers install the hardware and to install and configure the software.

- *Release Notes*—These notes will summarize the features of the product and detail any known bugs. (For subsequent releases, *Release Notes* will also document bugs fixed.)

For Integrators

We will provide the following single title for integrators:

- *Developing*—This is a guide that will detail our APIs for companies that want to integrate the Carambola 3000 with their products (for example, with factory automation software). This book will contain several example programs written in Java.

Customer Support

We will provide the following documentation for our own customer support organization:

- *Advanced Troubleshooting*—This guide will explain how to handle various potential customer complaints. Note that the *Installing* manual also contains a "Troubleshooting" chapter, but the *Advanced Troubleshooting* guide contains solutions (including remote debugging) that require more sophisticated knowledge of internals.

Although we will target *Release Notes* for consumers, we believe that customer support will also find them helpful.

Manufacturing

We will provide the following information for our manufacturing and procurement teams:

- *Schematics*—These will consist of highly detailed schematic diagrams; our manufacturing team requires these schematics to assemble the product.

- *Specs*—These documents contain tolerances and specifications for all hardware components; our procurement department requires these documents.

Media

We will ship documents in the formats listed in Table 3-2.

TABLE 3-2 How We Will Distribute the Documentation Set

Title Abbreviation	Medium	Distribution
Planning	PDF file	We will e-mail this file to consumers right after they buy the product.
Installing	Hard copy	We will kit these on the factory floor.
Release Notes	TXT file	This will be on the installation CD.
Developing	PDF file	This will be on the installation CD.
Advanced Troubleshooting	PDF file	We will post this on our corporate intranet.
Schematics	VSD files	We will post this on our corporate intranet.
Specs	PDF file	We will post this on our corporate intranet.

Issues

We've never used the print vendor before, and we're a little worried about quality.

Summary of Documentation Specifications

Before sending out your doc spec for review, ask yourself the following questions:

- Are the right people on your distribution list? Will anyone be offended if he or she is not on the distribution list? Conversely, do you have too many people on your distribution list? (If you put too many people on the list, then many readers will not bother reviewing the doc spec because they will assume that someone else will do it.)

- Is the schedule achievable? Can you really write a first draft that quickly? Have you ever written documents of this size before? If not, give yourself plenty of cushion—writing might take longer than you think. Have you allotted a suitable amount of time (not too long and not too short) for others to review this document?

- Does your outline contain all the topics that your target audience needs? Are the topics appropriate for your target audience, or are the topics too hard?

Before sending out your documentation project plan for review, ask yourself the following questions:

- Does your documentation set cover all audience segments?

- Do you have the right amount of information for each segment of the target audience? In other words, does the proposed documentation set meet the needs of all segments? Typically, the honest answer to this question is no. In this case, does the documentation set meet the needs of the primary segment?

- Can your team write all the specified documentation in the allotted time?

QUANTUM LEAP
When the engineering project changes, don't forget to update your documentation plans.

Writing: General Principles

This section explains the general principles of technical and scientific writing, from sections and paragraphs right down to sentences and individual words.

CHAPTER 4

Words

The writer's section of your local bookstore contains several lengthy books discussing proper word usage. The erudite writers of these books can spill tankers of ink explaining the many true meanings of the verb "to lie" or the creamy origins of the noun "seersucker." If you intend to get into this writing thing, reading one of these books will give you an appreciation for the subtleties of words and the obsessive requirements of the writer's work.

Why must you obsess about picking the right words in technical prose? In our litigious world, entire lawsuits hinge on improper word usage. Beyond crime and punishment, picking the wrong word in a manual can cost your enterprise a fortune in customer support. Conversely, picking the right word can unlock all sorts of new value in your inventions or ideas.

This chapter focuses on word choice in technical and scientific writing.

Jargon

Jargon is terminology used by experts for experts, that is, words and phrases that practitioners use to communicate within a particular field. When writing for other experts, you must use jargon. Without jargon, you'd be perpetually redefining ideas that your readers already understood. With jargon, you can just blurt out "capacitor," and your electrical engineer readership will know exactly what you're talking about.

The hard part to remember is that your lay audience does not speak jargon. (Hey, that's what makes them the lay audience in the first place.) Frequently, jargon is so ingrained in the way you speak and write that you no longer perceive jargon to be jargon. You may also find it hard to remember who your jargon-speaking peers are. For example, your fellow engineers all understand jargon, but will your manager understand the latest terms? Remember that most managers no longer perform feats of engineering every day, so although they'll understand a perennial like *capacitor*, they might no longer understand the latest terms.

Jargon Is a Dynamic Designation

Some words bob up and down between jargon and general use. Consider the word *transistor*. When transistors were first invented, this word was clearly electrical engineering jargon. Then, in the 1960s, cheap transistor radios flooded the American market, and the word *transistor* was on everyone's lips. Although lay people didn't know exactly what transistors did, they generally understood that transistors helped miniaturize gadgets. Since the 1980s, the term *transistor* has retreated into jargon, while solid-state terms like *CPU* and *RAM* have transitioned from jargon into everyday use. If you don't believe it, look at the computer ads in newspapers and lay magazines. These ads describe a CPU's clock speed and the amount of RAM without having to explain what these terms mean.

To define jargon for a lay audience, do either of the following:

- Define the term *in place*; that is, define the term when you use it for the first time.

- Defer the definition to a glossary.

For example, consider the following in-place definition:

Our tuner contains 50 *transistors*, which are tiny circuits that boost weak signals into strong signals. A transistor can turn a whisper into a scream.

Consistency

Renaming a part in midstream will drive your readers crazy; use the same name for the same part throughout the document. For example, if you refer to a certain part as a "widget" in Chapter 1, don't refer to that same part as a "gadget" in Chapter 2. Once a widget, always a widget. Remember that case is also part of a name. Therefore, you must be consistent in how you represent the case of a name. For example, a *widget* cannot suddenly become a *Widget* or (horrors!) a *WIDGET* because some readers will suspect that these are different parts.

It is even more confusing to give two different parts the same name. For example, never refer to one part as *the thingy* and then to a completely different part as *the thingy*. Overloading is useful in programming but awful in technical communication.

Many of you are probably thinking that the preceding two paragraphs are just common sense and that no engineer or technical writer would do such a thing. In fact, overloading and renaming errors are present in almost every technical manual. Oftentimes, it isn't the writer's fault—sometimes the fault lies with the people who originated the names in the first place.

Acronyms provide another prime opportunity for overloading, inconsistency, and confusion. For example, I once worked at a company that overloaded the acronym *PS* to symbolize four different products or technologies. As you can guess, an acronym should only symbolize *one* entity.

When new writers learn about the importance of variety in writing, they often oscillate between a technology's full name and its acronym. When you first introduce a term, you should also introduce its acronym, as in the following example:

> A Relational Database Management System (RDBMS) uses a table metaphor to store data in rows and columns.

After the initial use, you should pick one version—the more natural version—and use it exclusively. For example, the acronym *RDBMS* is far more natural than the expanded version, *Relational Database Management System*.

Verbs

When picking verbs, follow these guidelines:

- Choose strong verbs.

- Avoid overusing forms of *to be*.

- Provide variety in your choice of verbs.

- Use the same tense throughout the document.

Choose Strong Verbs

Strong verbs pack muscle and energy, moving the reader into action. Strong verbs are specific, precise, and dynamic. Weak verbs lie listless and dormant, generating boredom and sloth. Avoid the following famously weak verbs:

- to occur

- to happen

The preceding generic verbs don't convey anything specific—they just happen. Compare the following variants, noticing how much more powerful version 2 sounds than version 1:

1. If you spread organic compost and limestone, bigger vegetables will **occur**.

2. Spreading organic compost and limestone **generates** bigger vegetables.

Avoid Overusing *To Be*

Although the verb *to be* is fundamental and essential, you must not overuse it. Many writers reach for it automatically just because it is so handy. Note that *to be* simply exists; it does not describe.

Many writers team up *to be* with the generic noun *there* to produce *there is* or *there are*. This combination is very weak. For example, compare the following variants, noticing the stronger pull of version 2 over version 1:

1. **There is** a gravitational force that pulls on satellites.

2. **The Earth's** gravitational force pulls on satellites.

Version 2 replaces a generic subject (*There*) with a specific subject (*The Earth*). The weak verb (*is*) in version 1 disappears in version 2.

The wise writer keeps *there is* and *there are* to a minimum. When editing or revising, rewrite this construction into something more specific.

Vary Your Verbs

Some writers habitually use the same small set of verbs over and over again. To illustrate, consider the following passage:

> Figure 1 **illustrates** the molecular structure of ethyl alcohol, which is an intoxicant. Figure 2 **illustrates** the molecular structure of methanoate, which smells like raspberries. Figure 3 **illustrates** the molecular structure of...

The topic is tasty, but the verbs are monotonous. (Each sentence is also plagued by a monotonous grammatical format, but that's a different topic.) Instead of using *illustrates* three times in a row, try *specifies* or *shows* or *presents* or *exemplifies* or...

Keep Verbs in the Same Tense

You should maintain the same tense for all verbs within a paragraph. Paragraphs that mix tenses (like the following) sound amateurish:

> Hurricanes **require** high sea-surface temperatures. Another requirement **was** favorable high-level winds. With light winds aloft, hurricanes **will gain** power.

The present tense is usually the wisest choice for technical prose. However, the past tense is appropriate for portions of lab reports. After all, lab reports describe what has previously transpired.

You should generally maintain the same tense throughout an entire document. Some editors are extremely insistent on this point. I prefer a more pragmatic approach. For example, consider a book that describes previous research, present research, and potential future research. Clearly, such a book requires multiple tenses.

Adjectives and Adverbs

Good technical prose ladles in healthy doses of verbs and nouns and generally sprinkles in just an occasional dash of adjectives and adverbs. This is a real pity since adjectives and adverbs can spice a dull dish to life. Unfortunately though, marketing and advertising have made technical people suspicious of adjectives and adverbs. For example, consider the following misguided passage from a hardware manual:

> Our video board is extremely fast.

The phrase "extremely fast" smacks of marketing-speak, which raises doubts in engineers' minds. Once the seeds of doubt are planted, engineers will view the remainder of the text skeptically. When you are writing for engineers and scientists, it is much wiser to stick to objective, numerically based facts; for example:

> Our video board can render 1.2 million polygons per minute.

The preceding sentence presents an objective fact, which allows the reader to come to his or her own conclusions. Analytical people greatly prefer to come to their own conclusions. Let the mantra "Just the facts, ma'am" echo across your brain as you write.

Before going overboard, note that certain uses of adjectives are just fine. For example, using adjectives to describe physical appearance is essential. The following passage—from a bomb-defusal manual—would be catastrophic without adjectives (*red* and *blue*):

> Never cut the blue wire; cut the red wire only.

Similarly, occasional adverbs and adverb phrases are also fine, as long as they are objective and specific. For example, the adverb phrase (*70%–90% more quickly*) in the following sentence will not annoy technical readers:

> Adding a second CPU will make our software run 70%–90% more quickly.

Pronouns: *He, She,* and *They*

The pronouns *he* and *she* stir up plenty of trouble. Consider the following variants (all of which are aimed at someone other than the user):

1. When a user enters the password, they are redirected to the home page.

2. When a user enters the password, s/he is redirected to the home page.

3. When a user enters the password, he or she is redirected to the home page.

4. When a user enters the password, she is redirected to the home page.

5. After users enter the password, they are redirected to the home page.

6. After entering the password, the user is redirected to the home page.

Variant 1 is grammatically incorrect no matter how much we'd all like it to be proper. Unfortunately, in English, you cannot replace a singular *user* with a plural *they*.

Variants 2 and 3 are politically and grammatically correct but rather clumsy. In addition, *s/he* isn't actually a word, and it may confuse nonnative speakers.

Variant 4 is okay, but you must ensure that this particular user (the one entering the password) doesn't morph into *he* later on.

Version 5 is grammatically correct. After all, the plural *users* matches the plural pronoun (*they*). However, the sentence is not completely logical. According to the sentence, users enter the password. Literal-minded readers might imagine that multiple people must enter the same password. Furthermore, according to the sentence, all users will be taken to the home page. Again, literal-minded readers might imagine that all readers will be transported simultaneously to the home page, which is not the intent.

Version 6 is your best bet, basically because it skips right over the entire excruciating gender problem and replaces the pronoun with the androgynous *the user*.

In some situations, "going plural" (as in version 5) is advantageous, but rewriting sentences to avoid using pronouns (as in version 6) is always safe.

Pronouns: *You*

I love *you*. You should love *you*, too.

The second person plural is a valuable pronoun in technical communication. The pronoun *you* personalizes instruction. It sends a signal to the reader that says, I care for you. In addition, using the pronoun *you* improves an audience's attention. (Yeah, I'm talkin' to you.) For example, compare the following variants:

1. When the centrifuge starts, the operator may hear a brief cranking sound.

2. When the centrifuge starts, you may hear a brief cranking sound.

Doesn't version 2 sound more personal and friendly than version 1? In real life, wouldn't you prefer that someone refer to you as *you* rather than as *the operator*? Naturally, the writer must first be certain that you (the reader) really are the operator.

Imperative verbs are commands. Imperative verbs act as an implied *you*. Do not place *you* in front of an imperative verb. (In fact, when you place a pronoun in front of an imperative, the verb is no longer imperative.) For example, although the following two variants are synonymous, version 1 sounds more natural:

1. Start the centrifuge.

2. You start the centrifuge.

Pronouns: *It* and *They*

If used infrequently, the pronouns *it* and *they* are benign. However, many readers mistakenly use these pronouns with abandon. *It* becomes the favorite subject for lazy passages, used instead of a more specific noun. For example, notice how many times the following passage contains *it*:

> The carambola is a delicious fruit. **It** is native to Malaysia, but **it** now grows in various other places, as well. **It** is very sensitive to cold weather. **It** is often called "star fruit."

The following passage reduces occurrences of *it*:

> The carambola is a delicious fruit. Although native to Malaysia, this fruit now grows in various other places as well. The tree is very sensitive to cold weather. **It** is often called "star fruit."

If you look carefully at the two passages, you might spot another problem with *it*, which is that writers sometimes change the meaning of *it* within a paragraph. In the first passage, note that *it* refers to a fruit. However, a few sentences later, *it* has become a tree. In the second passage, notice that the tree—not the fruit—is very sensitive.

See if you can detect the transgression in the following passage:

> Oranges are higher in Vitamin C than avocados, but **they** have more protein.

Which has more Vitamin C: oranges or avocados? Readers cannot tell. A lazy writer has introduced two nouns and then mapped only one (and we cannot determine which one) to *they*. You must always be perfectly clear what *they* refers to.

Fluffy Phrases

One of the oldest cliches in writing is to avoid old cliches when writing. Unfortunately, due to habit, most educated people rely on stock, fluffy phrases when a single word will do. These phrases are so ingrained in your writing that you can probably no longer recognize them as fluffy. For example, in the following variants, notice that version 2 is more concise and clearer than version 1:

1. Failure to destroy Object Q will **cause the introduction of** memory leaks.

2. Failure to destroy Object Q will **introduce** memory leaks.

Version 2 eliminates three words from version 1.

Table 4-1 identifies a few common wasteful phrases. Once you start noticing these wasteful phrases, many more will pop at you.

TABLE 4-1   Replace Wasteful Phrases with Their Frugal Equivalents

Wasteful	Frugal
applies pressure	pushes
at the current time	now
causes the introduction of	introduces (or causes)
in spite of the fact that	although
develops a habit of	habitually
as well as	and
provides variety	varies
provides a detailed description of	details (v)
makes changes to	changes
provides some clarification of	clarifies
that is to say	that is

In some sentences, you can simply remove the wasteful phrase *that is to say* altogether rather than replacing it with *that is.*

Commonly Confused Words

This section describes the words most commonly confused in technical prose.

That and *Which*

Even professional writers often have trouble using the words *that* and *which* correctly, so pay careful attention to that which I'm about to say. The paradox is that although the two words are synonyms, you may not use them interchangably. Without getting into excruciating grammatical terms, the rule of thumb is as follows:

- Use *which* immediately after a comma.

- Do not use *that* immediately after a comma.

Compare the following sentences:

- Maple trees produce a sweet sap that farmers process into syrup.

- Maple trees produce a sweet sap, which farmers process into syrup.

Can and *May*

Writers chronically interchange *can* and *may*. Table 4-2 highlights the difference.

TABLE 4-2 *Can* versus *May*

Word	Meaning	Example
Can	Is able to	This computer can execute 20 billion instructions per second. (This computer is able to execute 20 billion instructions per second.)
May	Has permission to or is allowed to	If you have write permission on the directory, you may create a file in it. (If you have write permission on the directory, you are allowed to create a file in it.)

The word *may* also indicates "a possibility of." For example, in casual speech, the meaning of the following sentence is perfectly clear:

It may rain.

In formal technical prose, you should avoid using *may* to indicate possibility as it leads to confusion with the other common meaning of *may*. Within formal technical prose, it is wiser to use "possibility" or "probability" instead of *may*, as in the following example:

> The Weather Service forecasts a 50% probability of rain.

Effect versus Affect

Effect and *affect* are similar-sounding, common technical words whose meanings change depending on whether the words are being used as a noun or as a verb. Cutting to the chase, Tables 4-3 and 4-4 highlight their *most common uses* in technical and scientific writing.

TABLE 4-3 When Used as a Verb

Verb	Meaning	Example
To affect	To influence	Air pressure affects acoustics. (Air pressure influences acoustics.)
To effect	To bring into existence	April showers effect May flowers. (April showers give birth to May flowers.)

TABLE 4-4 When Used as a Noun

Noun	Meaning	Example
Affect	Emotional appearance (a term in psychology)	He had a flat affect. (He had a flat emotional appearance.)
Effect	The result or outcome	Add a catalyst to produce the desired effect. (Add a catalyst to produce the desired outcome.)

Its and It's

If the English language were an engineering project, I'd submit a high-priority change request to create a substitute word for *its*. Unlike every other possessive in English, *its* does not contain an apostrophe. Table 4-5 differentiates between *its* and *it's*.

TABLE 4-5 *Its* versus *It's*

Word	Meaning	Example
Its	Possessive (something belongs to *it*)	Its coat is gray.
It's	A contraction for *it is*	It's a beautiful day in the neighborhood.

Summary of Words

When reviewing your word choices, ask yourself the following questions:

- Are these words appropriate for the target audience? Does the target audience know what these words mean? Are any words too "big" for their own good?

- Does this document contain jargon? Does your target audience know the definition of this jargon? If your target audience is a lay audience, do you provide definitions?

- Are all the words clear? Can you replace a word with a more specific or accurate word?

- Are the verbs strong or generic? Can you rewrite sentences containing *there is* or *there are*?

- Are all the verbs in the same tense? In general, they should be.

- Does the same verb appear repeatedly? Vary your verbs.

- Are any words unnecessary? Can you reduce or remove certain words?

- Are any adjectives or adverbs unnecessary? When writing a technical document for a technical audience, you should remove most adjectives or adverbs that suggest marketing pressure.

- Are the pronouns grammatically correct? For example, do any sentences refer to one person as *they*?

- Does the document overuse *it*? Are any uses of *it* or *they* ambiguous? In other words, does the reader always know which noun *it* or *they* refers to?

- Does the document refer to the reader as *you*? It should.

- Does the document apply each word correctly? If in doubt, refer to a dictionary or style guide. Pay particular attention to *that* and *which*, *can* and *may*, and *affect* and *effect*.

CHAPTER 5

Sentences

A run-on is the kind of sentence that just keeps rolling along until the furthest shores of infinity, churning and churning without really saying anything, reiterating what has come before and not really advancing the point until the reader is bored and moves on to the next book—any book (even an art history book)—on the shelf at the bookstore. Yep, that's a run-on sentence. That is the kind of sentence that you are not going to write. That's because you, my friend, are an engineer or scientist, and people with logical minds hate fluff. You adore signal and despise noise.

Sentences don't have to be long to be noisy; sentences can be as short as Napoleon but still be noisy if they don't say anything valuable. Similarly, sentences can be filled with good data that gets lost in a confusing structure.

Writers generally cannot detect noise in their own prose. After all, noisy writing emits no sound or spark (unless you count the grinding of a frustrated reader's teeth). Fortunately, a good editor can detect noise in your sentences and help you filter it out. Like any engineering process, writing effective sentences must endure critical analysis and iteration.

This chapter takes you through the key components of writing good sentences, which include the following:

- using active voice
- keeping sentences short
- aiming for clarity above all

Active Voice and Passive Voice

In an **active-voice** sentence, the subject acts on the object. For example, the sentence in Figure 5-1 is in active voice because alpha particles (the subject) act on the nucleus (the object).

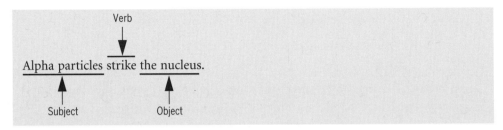

FIGURE 5-1 Active-Voice Sentence

In a **passive-voice** sentence, the subject is acted upon by the object. In most passive-voice sentences, the object appears at the beginning and the subject appears at the end. For example, the sentence in Figure 5-2 is in passive voice because alpha particles (the subject) are acted upon by the nucleus (the object).

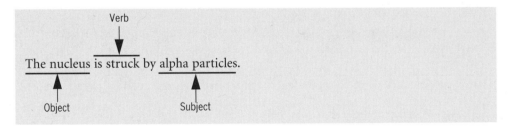

FIGURE 5-2 Passive voice sentence

Table 5-1 provides a few more examples. The final passive voice example omits a subject altogether. (Who measured barometric pressure? It is hard to say.)

TABLE 5-1 Examples of Active Voice and Passive Voice

Passive Voice	Active Voice
GIFs are displayed by the browser.	The browser displays GIFs.
Two parties are connected by telephones.	Telephones connect two parties.
Barometric pressure was measured.	The team measured barometric pressure.

Active Voice Is Better

Active-voice sentences offer the following compelling advantages over passive-voice sentences:

- Active-voice sentences are usually a little shorter than their passive-voice equivalents.

- Most readers find active voice clearer and more natural than passive voice.

- Active-voice sentences are more powerful and energetic than passive-voice sentences.

- Readers for whom English is a second language will find active voice easier to understand than passive voice. (Students of English as a second language invariably learn active voice long before they learn the passive voice.)

In summary, active-voice sentences are generally superior to passive-voice ones.

Well... How Did I Get Here?

The urge to write passive-voice sentences is often the by-product of a good education. Advanced academic writing is often riddled with passive voice. Passive voice generates an effete air, while active voice just sounds so plebeian. Class connotations aside, clarity is the shining gold standard of technical communications, and clarity requires active voice.

The worst misuse of passive voice is the sentence that omits a subject altogether. In technical and scientific writing, it is critical that the reader understand who is doing what to whom. By tradition, scientists often skip subjects in lab reports. According to this tradition, the absence of subjects makes lab procedures sound less subjective and more factual. Thus, temperatures are taken and slides are viewed. However, outside of Harry Potter, microscopes do not see; humans see through microscopes. It is high time to acknowledge the actor in each action. The following list provides a few convenient actors for lab reports:

- the authors

- the study

- the lab team

When Is Passive Voice Okay?

Using passive voice once in a while is okay. For example, you might prefer to use passive voice when you want to stress the object rather than the subject. For example, consider the following active-voice sentence:

> Hurricanes seldom hit New York City.

The preceding active-voice sentence is dandy if the topic is *hurricanes*. However, if the topic is *New York City*, then a passive-voice sentence such as the following might make more sense:

> New York City is seldom hit by hurricanes.

Sometimes you use passive voice to obscure the subject intentionally, perhaps to spare blame. For example, consider the following two politically expedient, passive-voice sentences:

> Mistakes were made.
>
> Hmm, the last autoclave has been taken.

You may sprinkle in an occasional passive-voice sentence for the sake of variety to break up the monotony of a string of active-voice sentences. (We'll revisit sentence variety in the next chapter.)

Short = Sweet

Long story short—readers prefer skinny sentences to fat ones. This is a tough lesson to learn because, as you got older, your teachers rewarded you for writing longer sentences. In fact, your ability to write bloated sentences was an indicator that you were "educated." Good engineers do not simply invent clever devices. Rather, they invent devices that can be manufactured as cheaply as possible. Your goal as a writer is similar—you must convey information in a minimum of words.

Long sentences are evil for the following reasons:

- They are visually intimidating.
- They are harder to understand than short ones.
- They often obscure the writer's intent.

In Table 5-2, note the relative clarity of the shorter version.

TABLE 5-2   The Long and Short of It

Long Version	Short Version
The efficiency of Web browsers in displaying graphic images is in direct correlation to the disk resources required by the aforementioned graphic images.	Web browsers display small graphics files more quickly than big graphics files.
It would be extremely tedious, and in most cases irrelevant, for us to concern ourselves with the atomic or molecular structure of macroscopic objects whose gravitational attraction is to be studied when the object's mass is essential.	Gravitational force depends on an object's mass, not on its chemical structure.
Liquid precipitation on the western Iberian peninsula is primarily deposited on flat, non-mountainous regions.	The rain in Spain falls mainly on the plain.

Keeping It Bite-Size

At Chinese restaurants, the chef chops food; patrons do not. This is why you will rarely find knives on the table at Chinese restaurants. As a writer, you must become the Chinese chef—chopping your sentences into delectable morsels so that your readers do not have to work so hard.

Causes of Long Sentences

I've rounded up the usual suspects for killing your readers' time:

- jamming two or three ideas into a single sentence

- using passive voice

- being repetitious

- being repetitious

- being repetitious

- using words or phrases that do not pull their weight

- embedding lists that should be broken into bulleted or numbered lists

When is a sentence too long? It is difficult to come up with a magic formula. A 12-word sentence that minces words might be too long, while a crystal clear 20-word sentence is sometimes just perfect. In general, though, you should rarely go beyond 25 words and almost never beyond 30 words.

Readability Formulas

Plugging prose into a readability formula generates a **readability quotient**. Most readability quotients are pegged to an educational level. Thus, a certain readability formula might indicate that a particular book is appropriate for eighth graders. Quite a few readability formulas are currently in use. Most rely on some mix of the following parameters:

- average number of words per sentence

- average number of "hard" words per sentence

- average number of characters or syllables per word

Unfortunately, many writers take readability quotients far too seriously, forgetting that these formulas omit common sense. For example, Ernest Hemingway wrote beautifully short sentences comprising short, simple words. Plugging some of his novels about bullfighting, combat, and sexual dysfunction through certain readability formulas yields books with a readability quotient suitable for fifth graders.

When writing technical prose for adult professionals, do not intentionally aim for a college-level readability quotient. Aim to make your message clear. Shoot for the best possible signal:noise ratio.

One Sentence = One Thought

A sentence should represent a single thought or idea. When a sentence holds multiple thoughts, you should pick one of the following tactics:

- Divide the sentence into multiple sentences.

- Divide the sentence into two parts connected by a semicolon.

Consider the following heavy sentence, pregnant with multiple ideas:

> Our sun is a yellow dwarf star, which is powered by nuclear fusion in a process that should last for a few billion more years or about as long as the balance is maintained between readily fusionable material and gravity, at which time it will transition to another star category.

The preceding sentence is a classic 50-word run-on, slogging along in a marathon when a series of short sprints would have worked better. The sentence contains four discrete thoughts (category, power source, duration, transition), so consider the following replacement containing four sentences:

> Our sun is a yellow dwarf star. Fusion powers all yellow dwarves. Our sun has enough fuel to last a few billion more years. When gravity can no longer keep atoms close enough to fuse, our sun will transition to another star category.

The resulting series of sentences is an improvement; however, it is rather choppy. To eliminate some of the choppiness, you might rewrite the opening as follows:

> Our sun is a yellow dwarf star, which is powered by fusion.

However, I'm not wild about this solution because this single sentence contains two discrete thoughts (category and power source). In addition, putting the power source after the comma relegates it to a second-class factoid. To work around these problems, connect the two thoughts with a semicolon instead of a comma, as in the following example:

> Our sun is a yellow dwarf star; fusion powers all yellow dwarves.

Notice that the power source is now on an equal footing with the category.

Parenthetical Clauses

A **parenthetical clause** is a digression, example, or elaboration within a sentence that is bounded by a pair of any of the following punctuation marks:

- commas

- parentheses

- dashes

For example, the following parenthetical clause is a digression bounded by a pair of dashes:

> Carbon—an element in all amino acids—can easily bond with hydrogen

If you eliminate a parenthetical clause, the resulting sentence must still make sense. For example, eliminating the preceding parenthetical clause leaves the following simple sentence:

> Carbon can easily bond with hydrogen.

The fiction writer Henry James loved parenthetical clauses and wrote enormous sentences, each of which would take well-educated Victorians an entire evening to read. Unfortunately, today's technical reader is not quite as patient. In technical prose, occasional parenthetical clauses are fine, but don't overuse them. No sentence in technical prose should contain more than one parenthetical clause.

In many programming languages, it is fine to place a pair of parentheses inside another pair of parentheses. However, you should avoid this practice in technical prose. Humans are less comfortable with recursive parsing than compilers.

Of course, commas and dashes can mark more than just parenthetical clauses. Both can also denote simple pauses. Since frequent pausing doesn't appeal to technical readers, you should not overuse commas and dashes.

Summary of Sentences

When reviewing your writing, ask yourself the following questions about each sentence:

- Is this sentence clear? Will the target reader understand it? Is this sentence too complex for the target reader, or does it require knowledge that the target reader does not possess? Conversely, does this sentence insult the target reader's intelligence? All other items in this checklist are subservient to clarity; make clarity your primary goal.

- Is this sentence really a sentence? (All sentences must contain a verb.)

- Is the verb strong and specific, or is it weak and generic? The verb is generally the most important word in the sentence—choose a powerful one.

- Does this sentence contain an embedded list? If so, you should typically break it out to form a bulleted or numbered list.

- Is this sentence in passive voice? Most sentences should be in active voice.

- How many ideas does this sentence convey? Each sentence should convey exactly one idea, although you can connect two closely related ideas with a semicolon.

- Is this sentence a run-on? Does it go yackety-yack and waste your time of day? If so, find a way to reduce its length, possibly by dividing it into two or more sentences.

- Is this sentence—while not a run-on—still longer than it needs to be? Always search for fat to trim.

- Does this sentence contain too many parenthetical clauses? Occasional parenthetical clauses are okay, but when overused they become literary speed bumps.

- **Do you need this sentence at all?** If you deleted this sentence, would the reader miss it? If the reader would not, erase it. After all, the reader will never know what you removed.

CHAPTER 6

Paragraphs and Sections

Somewhere in your schooling, a good English teacher probably taught you how to write good paragraphs. The key, the teacher said, was starting with a solid topic sentence that introduced the paragraph and provided its theme. The essays you wrote in high school contained long paragraphs, which really did benefit from topic sentences.

Technical writing differs from the literary essays you wrote in high school. Technical prose simply cannot afford a topic sentence for every paragraph. The paragraphs in technical prose are more utilitarian, less rigidly structured, and generally shorter than in literary criticism. In technical prose, two-sentence paragraphs are common. In such short paragraphs, a topic sentence is hardly warranted. The opening sentence in a technical paragraph does not need to state a theme; it merely needs to introduce the topic at hand or to build on the topic that preceded it.

A **section** is a collection of one or more paragraphs (plus any lists, tables, or figures). A section that consists of several paragraphs does require an introductory paragraph. This introductory paragraph should start with one of the following:

- a topic sentence
- a bulleted or numbered list of topics covered in this section

This chapter details the special requirements for paragraphs in technical prose.

Sentence Transitions

A **transition** is a word or phrase that helps ease readers into the next sentence. The following is a list of common transition terms in technical prose:

- however
- for example
- nevertheless
- by contrast
- by comparison
- in other words
- unfortunately
- that is (but don't use this one too often)

Transitions terms typically appear at the beginning of a sentence. For example, consider the following use of a transition term in the second sentence:

> Transitions terms typically appear at the beginning of a sentence. For example, consider the use of a transition term in this sentence.

When a sentence twists abruptly from the previous sentence, a transition term acts much like a squiggly road sign warning readers of a quick turn up ahead. For example, in the following passage, consider the sharp turn the second sentence makes from the first:

> Newton's formulas work remarkably well. These formulas fail miserably when objects are very small or are moving very rapidly.

Adding a transition term supplies a warning to the reader, so the following passage reads more easily:

> Newton's formulas work remarkably well. However, these formulas fail miserably when objects are very small or are moving very rapidly.

The wise writer pays attention to transitions. Without transitions, paragraphs often sound static and forced. With well-chosen transitions, writing sounds more natural and conversational. Ironically, speakers in actual conversations do not use many transition words.

Vocal inflections, facial expressions, and hand gestures act as transitions in real-life conversations.

In Table 6-1, notice how much more fluid the passages with transitions sound.

TABLE 6-1 The Value of Transitions

Without Transitions	With Transitions
The C compiler can process 1,000 lines of code per second. The rate slows when the host is fully loaded.	The C compiler can process 1,000 lines of code per second. **However**, the rate slows when the host is fully loaded.
Most commercial airplanes can cruise at altitudes up to approximately seven miles. The Boeing 757 cruises at 37,000 feet.	Most commercial airplanes can cruise at altitudes up to approximately seven miles. **For example**, the Boeing 757 cruises at 37,000 feet.

The only downside to using transitions is that they make sentences a little longer. Nevertheless, transitions aid clarity and help engage the audience.

Conjunctions Are Not True Transitions

It is tempting to start a sentence with *But* or *And*. But don't do it in technical writing. And here's why—*but* and *and* are conjunctions, great for intrasentence transitions but not appropriate for intersentence transitions.

Note that starting a sentence with a conjunction is fine in fiction. The rules of fiction are not as rigid as those for technical prose. Furthermore, in real-world conversations (which is something that fiction writers try to emulate), speakers start sentences with conjunctions all the time.

Although transitions are wonderful, don't feel obligated to force them into every sentence. Moderation is also powerful.

Paragraph Length

In high school English class, you may have been taught to write paragraphs to the following formula:

- The first (or topic) sentence introduces the paragraph.

- The middle three sentences are the body of the paragraph.

- The final sentence summarizes the paragraph.

Consequently, *every* paragraph consumed roughly five sentences.

When writing technical prose as an adult, you might have brought along that same rigid formula (one paragraph ~ five sentences). Allow me to liberate you—a paragraph should be as long or short as is needed. Some paragraphs should weigh a skimpy two or three sentences, while others should weigh a robust seven or eight sentences. Both weights are equally healthy.

Many good writers vary the length of paragraphs. Changing the length keeps readers awake. Some writers intentionally pop in an occasional lengthy paragraph after a quick series of short paragraphs. Your high school English teacher might not approve, but your readers will.

Yes, an occasional one-sentence paragraph is fine.

Beware of the rambling, massive paragraph that just never ends. When editing such a paragraph, I sense that the writer hasn't figured out the right key to press to trigger a new paragraph. A one-ton paragraph possesses an intimidating amount of girth and heft. Reading such paragraphs is a chore, which is why many readers simply skip over them. The obvious solution is to divide morbidly obese paragraphs into multiple svelte, healthy ones.

When writing for a medium (such as newspapers) that contains very narrow columns, your paragraphs must be exceedingly short, oftentimes only a sentence or two.

Paragraph Transitions

The opening sentence of a paragraph should flow fairly naturally from the preceding paragraph. For example, in the following passage, notice that the opening sentence of the second paragraph transitions by tightly summarizing earlier facts before segueing to the next topic:

> Dolphins are outstanding swimmers that can move at up to 30 miles per hour. Some dolphin species can hold their breath for up to ten minutes. Eventually though, dolphins must come up for fresh air since they have lungs, not gills.
>
> **Although dolphins both breathe air and swim expertly, dolphins are not amphibians.** True amphibians start their life in water prior to transitioning...

Many editors frown on starting a paragraph with a transition term, feeling that transition terms connect sentences, not paragraphs. For example, some editors would forbid the following paragraph opening:

> However, dolphins are not amphibians.

If you cannot make a paragraph flow naturally from its predecessor, then place that paragraph in a new or different section.

Sections

You should divide any document longer than a page or two into distinct sections. Doing so makes the document easier to write and easier to read. Each section must begin with a section **header**, which is its title. The following are some guidelines for sections and section headers:

- The opening sentence of a section should establish the section's purpose.

- Each section must contain at least one sentence. That is, you may not place two sections in a row without intervening text.

- Section names should be grammatically parallel. For details on parallelism, see "Parallel Lists" on page 70.

- Section names are not body text and should not be treated as such. For example, a section header cannot act as a valid introduction to a list; you must supply the introductory sentence in body text.

- Section levels should rarely go more than three levels deep, counting the title of the document (or the chapter title) as the top level. In other words, beyond that top-level header, you should only supply two additional levels of headers. At the fourth level, many readers lose track of the hierarchy, which leads to confusion.

Should You Number Section Headers?

Many documents provide a multilevel section number in front of each section name, as in the following section header:

3.6.4 Orcas Make Poor Pets

You should place section numbers on the following types of documents:

- **Engineering documents that serve a legal purpose.** For example, documents that are attachments to a contract almost always contain section numbers. Section numbers reduce ambiguity when lawyers discuss the contract.

- **Engineering documents in which section numbers are a contractual requirement**. For example, most military organizations require documents to contain section numbers.

When writing for a lay audience, avoid section numbers. To members of a lay audience, section numbers appear formal and intimidating, rather like a big red "nerd crossing ahead" warning flag being waved in their faces.

Summary of Paragraphs and Sections

When reviewing your writing, ask yourself the following questions about each paragraph:

- Does each sentence flow naturally into the sentence that follows?

- Do sentences sound choppy? If so, consider adding transition terms to the beginning of choppy sentences.

- Do any sentences start with conjunctions instead of proper transition terms?

- Does each paragraph flow naturally into the paragraph that follows?

- Do any paragraphs start with transition terms? They usually should not.

- Are any paragraphs too long? Be wary of paragraphs longer than eight sentences or so. Chop lengthy paragraphs into two or three shorter ones.

- Do all the sentences in a paragraph belong together, or do some sentences veer so far off course that they belong in another paragraph?

When reviewing your writing, ask yourself the following questions about each section:

- Is each section header descriptive? In other words, does each section actually describe what the section header says it should?

- Can any section headers be more specific? For example, calling a section header "Orcas" might not be not nearly as informative to readers as calling it "Orcas: Diet."

- Are all the paragraphs in a section related, or do some paragraphs belong in a new or different section?

CHAPTER 7

Lists

An offbeat publication from the 1970s, *The Book of Lists* by David Wallechinsky and Amy Wallace, was literally just a bunch of lists. However, readers loved it and made it into a best-seller. What could account for this book's popularity? Perhaps it was the titillating list categories, such as the top ten sex positions or ten worst dictators. Alternatively, perhaps readers just get a big kick out of lists. After all, lists neaten and straighten a ragged world. Lists condense information and force readers to focus. Even the layout of lists is favorable, making their information hard to ignore.

This chapter explores the following two kinds of lists:

- bulleted lists, which present unordered information
- numbered lists, which present ordered information

Inexperienced writers often confuse the two kinds of lists and pick the wrong one. Fortunately, the following rule of thumb will allow you to pick correctly:

> If rearranging the elements in a list would not alter the list's meaning, then the list should be bulleted. If rearranging the elements in a list would cause the list to become confusing or incorrect, then this should be a numbered list.

Bulleted Lists

A *bulleted list* is the smart alternative to an embedded list. For example, the following sentence contains an embedded list:

> A tropical cyclone consists of a large area of low pressure, a closed isobar, sustained winds greater than 25 knots, consistent thunderstorms around the center, and a warm core.

Elements within an embedded list tend to hide. Transforming an embedded list into a bulleted list provides far more clarity:

> A tropical cyclone consists of the following:
>
> - a large area of low pressure
> - a closed isobar
> - sustained winds greater than 25 knots
> - consistent thunderstorms around the center
> - a warm core

Readers enjoy bulleted lists, so use them liberally. Do not, however, *overuse* them—too many bulleted lists will transform your prose into a PowerPoint presentation.

Preface each item in a bulleted list with a bullet. A *bullet* is traditionally the following filled circle punctuation mark:

-

However, you could use another mark for the bullet as long as you do so consistently across the document.

How Many Elements Comprise a Bulleted List?

What is the minimum number of elements one can put in a bulleted list? Some writers say two elements and some say three. I feel that two elements is the usual limit. However, you may create a one-item bulleted list in documents that have conditioned readers to expect certain kinds of information in bulleted lists. For example, suppose each page of a document starts with a bulleted list enumerating achievements in pharmacology for a given year. If 1971 produced only one significant achievement, then the 1971 page should still begin with a one-element list.

Elements in Bulleted Lists

The text for each element in a bulleted list can be just about anything, as long as each element is grammatically consistent. (For more on consistency, see "Parallel Lists" on page 70.) In the following example, each element of the list is a singular noun:

> The following are simple trigonometric functions:
>
> - sine
>
> - cosine
>
> - tangent

List elements may also be sentence fragments (as in the first example on the preceding page) or complete sentences, as in the following example:

> The strength of the two primary physical forces depends on the following:
>
> - Gravitational force is proportional to the mass of the two objects.
> - Electrical force is proportional to the charge on the two objects.

Each element of a bulleted list can itself contain a sublist. Each element of a sublist must also start with a bullet. The bullet for the sublist must be indented further than the bullet for the main list. A main list and sublist provides an effective means for creating a hierarchy. A sublist can also provide details about entries in the main list; for example:

> The genus Canis includes the following members:
>
> - Wolves:
>   - Weight is between 20 and 80 Kg.
>   - Tail is often horizontal.
> - Coyotes:
>   - Weight is usually less than 20 Kg.
>   - Tail is usually down.

The Length of Each Element

Keep list elements short. In general, do not let list elements exceed three sentences. If you are producing lists for a medium with very narrow columns (such as a newspaper), then you should keep list elements down to a single sentence.

When list elements get too long, consider dividing them into two or more separate elements or possibly into a new section of your document, complete with subheads. For example, consider a bulleted list having the following structure:

Intro:

- Sentence 1. Sentence 2. Sentence 3.

  Sentence 4. Sentence 5.

- Sentence 6. Sentence 7. Sentence 8.

  Sentence 9. Sentence 10. Sentence 11.

- Sentence 12. Sentence 13. Sentence 14. Sentence 15.

The elements in the preceding list are too long to be effective as a bulleted list. Convert the preceding into a short bulleted list and three subheads as follows:

Intro:

- Sentence 1.
- Sentence 6.
- Sentence 12.

Subhead

Sentence 1 altered. Sentence 2. Sentence 3.

Sentence 4. Sentence 5.

Subhead

Sentence 6 altered. Sentence 7. Sentence 8.

Sentence 9. Sentence 10. Sentence 11.

Subhead

Sentence 12 altered. Sentence 13. Sentence 14. Sentence 15.

Numbered Lists

Numbered lists describe events that must happen in a certain order. Use numbered lists to describe a sequence of steps. For example, consider the following numbered list:

Perform the following steps to begin driving:

1. Turn on the ignition.
2. Release the emergency brake.
3. Put the car in drive.
4. Press the accelerator gently.

Changing the order of the elements in the preceding list would lead to car repairs. Therefore, the preceding list must be numbered; it cannot be bulleted.

QUANTUM LEAP:
In the preceding list, what do the first words (Turn, Release, Put, Press) in each sentence have in common? Each of these words is an imperative verb, meaning that it is a command for the reader to follow. It is a great idea to start each element of a numbered list with an imperative verb. The imperative will keep your sentences short and crisp.

Each element in a numbered list should describe one and only one action. For example, consider the following alteration:

Perform the following steps to begin driving:

1. Turn on the ignition and release the emergency brake.
2. Put the car in drive and press the accelerator gently.

At first glance, the preceding list looks pretty good. However, readers are more likely to miss details in the two-step list than in the four-step list. The wise writer does not place multiple actions in a single numbered step.

Do not start a numbered list element with the word *then*. The number prefacing each list element is an implied *then*.

Directions

Numbered lists provide the best medium for giving directions. For example, consider the following directions:

> To travel from my house to the college, follow these steps:
>
> 1. Travel 2.4 miles west on Occidental Blvd.
> 2. Turn right on Nordic Ave.
> 3. Travel 1.2 miles on Nordic Ave.
> 4. Turn left on Waverley Blvd.
> 5. Travel 0.6 miles.
> 6. Turn right into "Parking Lot C" and find a parking spot.

Avoid obfuscating your directions with too many extraneous details, such as "if you see Harry's Discount Law and Garden, you've gone too far." Nevertheless, consider providing a few helpful details to work through difficult patches; for example, the following revision to step 4 provides some useful troubleshooting information:

> 4. Turn left on Waverley Blvd. The street sign is missing, so look for Puddingstone Bank on the corner of Waverley.

Avoid ordinal numbers in directions. Do not, for example, say that Waverley Blvd. is at the third traffic light. Ordinal numbers produce a surprising amount of confusion. (*When should we have started counting traffic lights? Did the flashing yellow light at the crosswalk count as a traffic light?*) Ordinal numbers also make troubleshooting difficult.

Know your audience. Are your readers mathematically inclined and content to deal with mileage figures and odometers? Are your readers more word oriented, preferring textual descriptions of hills and pretty red houses on the corner?

As you write a numbered list, imagine that readers are checking off each step as they complete it. Keep each entry discrete.

Introductions to Lists

You may not just blurt out lists; you must introduce them properly. Introductions are more than grammatical etiquette; they are essential for meaning.

> • carambolas
>
> • milk chocolate
>
> • corn chips

Whoa—did you see how I just blurted out the preceding list? What did those items mean? Since I didn't introduce the list properly, you probably have no idea. I didn't supply enough context to help you comprehend the list.[1]

When introducing a list, try to work the word *following* into the sentence. For example, consider the following beautiful introductory sentence:

> The following is a list of deciduous hardwoods:
>
> • oak
>
> • maple

Alternatively, consider the following variants on the preceding introduction:

> Deciduous hardwoods include the following examples:
>
> Examples of deciduous hardwoods include the following:

By putting *following* in an introductory sentence, you are likely to produce a useful context for the list. Don't try to understand this magic—just accept it.

Look Out Below

Never use the word *below* when introducing a list. Lists have a nasty habit of sliding to the next physical page. Thus, the list might not really be spatially below the sentence that introduces it. Using the word *following* instead of *below* keeps you safe, even if the list moves to the top of the next page. For the same reason, avoid the word *above* when referring to a previous list. Use the word *preceding* instead.

---

1. These are my favorite snack foods.

Parallel Lists

Each element in a list must have the same grammatical form. For example, if the first element in a list is a singular noun, then all elements must be singular nouns. If the first element is a complete sentence, then all elements must be complete sentences. If the first element in a list starts with an imperative verb, then all elements must start with a imperative verb. In addition, punctuation and case must be consistent for all elements in a list.

A list in which each element is consistent is said to be *parallel,* and a list of inconsistent elements is called *nonparallel.* Creating parallel lists does not come naturally to most writers—it requires a fair amount of discipline and a frustrating amount of rephrasing. Nevertheless, one of the surest ways to produce an amateurish document is to pepper it with nonparallel lists.

Beyond simple grammar, each element in a parallel list must *logically* belong to that list.

On this page and the next, we will consider a few parallel and nonparallel lists. See if you can determine which lists are nonparallel and why. (Note that the answers follow the sample lists.) We begin with the following list:

- lemon
- lime
- orange

The preceding list is beautifully parallel. All the elements are singular nouns.

- lemons
- lime
- orange

The preceding list is nonparallel because the first element is plural, while the other two elements are singular.

- Lemon
- Lime
- orange

The preceding list is nonparallel because, unlike the first two elements, *orange* did not start with an uppercase letter.

- lemon
- lime
- orange tree

At a grammatical level, the preceding list is okay; however, this is a nonparallel list because the third entry just does not logically belong with the other two.

1. Insert the CD.
2. Invoke `setup.exe`.
3. Answer the prompts.

The preceding list is parallel. All three elements are simple verb phrases. Each of the phrases starts with an imperative verb. All end with a period (as they should).

1. Insert the CD.
2. Invoke `setup.exe`.
3. Answer the prompts.

The preceding list is nonparallel. Notice that the first two elements end in a period, but the third element does not. To create a parallel list, you must use punctuation consistently. By the way, most writers place a period at the end of list elements that are verb phrases or complete sentences, and they omit the period from list elements that do not contain a verb.

1. Insert the CD.
2. Invoke setup.exe.
3. You must answer the prompts.

The preceding list is nonparallel because, unlike the first two elements, the third element starts with a pronoun.

Capitals and Periods in List Elements

Punctuate each element of a numbered list as you would a sentence; that is, capitalize the first letter of the opening word and end the element with a period. For a bulleted list, if the list elements are sentences or verb phrases, supply sentence punctuation. Otherwise, do not supply sentence punctuation.

Summary of Lists

When reviewing your writing, ask yourself the following questions about each list:

- Should this list be bulleted or numbered?

- Does my text contain any embedded lists? Should these embedded lists be broken out into bulleted or numbered lists?

- Should any complex list elements be converted into sublists?

- In a numbered list, do my elements begin with an imperative verb? (Numbered lists frequently begin with an imperative, but they don't have to.)

- Is the first element in a numbered list labeled as number 1?

- Is each list properly introduced? Does the introductory sentence end with a colon? Does any text refer to list *below* or *above* instead of using the proper words *following* or *preceding*?

- Are all list elements parallel?

CHAPTER 8

Tables

T ables are a terrific organizational tool for communicating certain kinds of data. They provide a pleasant visual break from page after page of prose. They divide an untidy world into nice neat fields, appealing to the obsessive-compulsive in us all. When organized appropriately, they act as an outstanding reference medium, allowing readers to quickly find what they seek.

Technical readers generally adore tables. Most lay readers appreciate simple, well-organized tables but often shy away from complex tables containing too many columns. Many technophobes are afraid of tables simply because they look rather technical.

Good tables have the following characteristics:

- They begin with a good caption that summarizes the contents of the table.

- Every column contains a clear, accurate header.

- The contents within each column are grammatically parallel.

- They are organized so that readers can easily find what they seek.

Bad tables often exhibit the following two problems:

- The contents of cells are so concise that they become cryptic.

- The contents of cells are so long-winded that the concise value of tables is lost.

Column Headers

You should provide headers for every column in almost every table. Good headers make data much easier to understand. However, writers often omit column headers, believing that the contents of each column are so obvious that headers are a waste of space. In fact, bad or missing headers can lead to ambiguity and miscommunication. For example, consider Table 8-1, which contains no column headers.

TABLE 8-1 Members of the Willow Family (without headers)

Crack Willow	Oblong lance	Europe and Asia
Peach-leaved Willow	Lance or ovate-lance	North America
Purple Willow	Parallel sides	Europe, Asia, and Africa
Weeping Willow	Lance to linear	Asia

Although the contents of the leftmost column of Table 8-1 seem fairly obvious, the values in the other two columns could have several possible meanings. Table 8-2 provides column headers, which clarify matters considerably.

TABLE 8-2 Members of the Willow Family (with headers)

Species	Shape of Leaves	Continent(s) of Origin
Crack Willow	Oblong lance	Europe and Asia
Peach-leaved Willow	Lance or ovate-lance	North America
Purple Willow	Parallel sides	Europe, Asia, and Africa
Weeping Willow	Lance to linear	Asia

When naming columns, be as specific as possible, even if the resulting name is not concise. For example, in the previous table, although *Continent* is more concise, *Continent of Origin* is far less ambiguous.

Units of Measure

As someone with scientific training, you know how important it is to specify accurate units of measurement. Technical writers specify table units in one of the following three places:

- in each cell
- in table footnotes
- in column headers

Specifying the unit in each cell provides unambiguous data. However, it also generates redundant clutter, as in Table 8-3.

TABLE 8-3   Specifying Units in Each Cell

Species	Height	Trunk Diameter	Leaf Length	Fruit Length
Black Walnut	70–90 feet	2–3 feet	12–24 inches	1.5–2 inches
Pawpaw	40 feet	1–1.5 feet	6–12 inches	4 inches
Red Mulberry	25–40 feet	1–1.5 feet	3–5 inches	1 inch

Providing footnotes reduces clutter but also reduces clarity; readers must move their eyes to multiple spots to get the full story. When viewing Table 8-4, notice how many turns your eyes must take to locate the trunk diameter of a black walnut.

TABLE 8-4   Specifying Units in Table Footnotes

Species	Height[a]	Trunk Diameter[a]	Leaf Length[b]	Fruit Length[b]
Black Walnut	70–90	2–3	12–24	1.5–2
Pawpaw	40	1–1.5	6–12	4
Red Mulberry	25–40	1–1.5	3–5	1

a. In feet
b. In inches

Specifying units in column headers, as in Table 8-5, is usually the best approach.

TABLE 8-5   Specifying Units In Each Column Header

Species	Height (in Feet)	Trunk Diameter (in Feet)	Leaf Length (in Inches)	Fruit Length (in Inches)
Black Walnut	70–90	2–3	12–24	1.5–2
Pawpaw	40	1–1.5	6–12	4
Red Mulberry	25–40	1–1.5	3–5	1

Referring to Tables

In text, refer to a table as *the following table*, *the preceding table*, or as a specific table number (*Table 8-12*). Don't refer to the table by its caption (*the Mulberry table*).

Arrangement of Columns and Rows

For each table, you must pick one column to be the central organizing unit. Place this primary column in the leftmost position in the table, primarily because English readers are accustomed to starting from the left and moving to the right.

Compare Tables 8-6 and 8-7.

TABLE 8-6 North American Tree Families

Family	Species	Height (in Feet)
Bean	Honeylocust	70–80
	Kentucky Coffee Tree	60–80
Beech	Chestnut Oak	40–60
	White Oak	80–100

TABLE 8-7 North American Tree Species

Species	Family	Height (in Feet)
Chestnut Oak	Bean	40–60
Honeylocust	Beech	70–80
Kentucky Coffee Tree	Bean	60–80
White Oak	Beech	80–100

Choose Table 8-6 when readers are more familiar with tree families and Table 8-7 when readers are more likely to look up entries by species.

Tables in Documents Are Not Tables in Relational Databases

Tables in documents are not subject to the same rules as tables in relational databases. For example, each table in a relational database must define one column to be the primary key. The value of a primary key must be unique. In other words, two rows in a relational database table may not share the same primary key value. It is tempting to believe that the leftmost column in a document table is a primary key, but as Table 8-6 illustrates, the leftmost column does not have to be unique.

In some cases, two columns may appear equally qualified to serve as the leftmost column. When this happens, the best solution is to present the same information in two separate tables. For example, suppose you are writing a document that compares UNIX and DOS. Further suppose that some members of the audience are more familiar with UNIX and others with DOS. Instead of flipping a coin, you should provide two tables like Tables 8-8 and 8-9.

TABLE 8-8   DOS Commands if You Already Know UNIX

UNIX Command	Equivalent DOS Command
cd	CHDIR or CD
ls	DIR
mv	RENAME
rm	DEL

TABLE 8-9   UNIX Commands if You Already Know DOS

DOS Command	Equivalent UNIX Command
CHDIR or CD	cd
DEL	rm
DIR	ls
RENAME	mv

Readers generally prefer alphabetical order over any other categorization scheme. Writers, for some strange reason, often organize tables using nonalphabetic schemes, generally because other schemes feel more "natural" to them. When a table isn't in alphabetical order, readers often get annoyed, trying to figure out the writer's scheme. When you have the luxury of time, consider providing the same information in two tables—one organized alphabetically and the other organized by some other organizational scheme. Alternatively, consider breaking a large table into a series of smaller tables, each representing a different category. Then, sort alphabetically within each small table.

Parallelism in Tables

"Parallel Lists" on page 70 describes the importance of parallelism in lists. In fact, this same concept is also important in tables. In particular, keep the text within all cells of a column parallel. To explore this idea, consider Table 8-10, in which every column is nonparallel.

TABLE 8-10 Characteristics of Tree Families (nonparallel columns)

Family	Fruit	Comments
Beech	Edible nut	Needs a wet environment.
Maples	Double-winged	The wood is hard.
Pine	A wood cone	Produce turpentine and rosin

In the preceding table, note the following problems:

- In the *Family* column, the first two entries are singular, but the last entry is plural.

- In the *Fruit* column, the first entry consists of an adjective and a noun. To be parallel, the second and third entries must contain one or more adjectives followed by a noun.

- In the *Comments* column, the first entry is a sentence that starts with a verb; therefore, the second and third entries should have the same form. Furthermore, since the first entry ends with a period, subsequent entries must end with a period. The phrase *Produce turpentine and rosin* isn't grammatically correct.

Table 8-11 shows a corrected, parallel version of Table 8-10.

TABLE 8-11 Characteristics of Tree Families (parallel columns)

Family	Fruit	Comments
Beech	Edible nut	Needs a wet environment
Maple	Double-winged fruit	Grows hard wood
Pine	Wood cone	Produces turpentine and rosin

Although tables must be parallel *within* columns; they do not have to be parallel *between* columns. For example, it is okay that all cells in the *Family* column are nouns, but all entries in the *Comments* column are verb phrases. Nevertheless, it is still a good idea for all entries throughout a table to be consistent in number (singular versus plural) and in tense.

Amount of Text in Cells

Text in tables should be clear and highly concise. Never get long-winded inside a cell. For example, Table 8-12 contains too much text in each cell.

TABLE 8-12 Characteristics of the Mulberry Family

Species	Fruit	Comments
Osage Orange	The fruit has a bizarre, brainlike exterior. It is about the size of a citrus orange.	The tree has scary-looking spikes. When several trees are planted close together, the hedge forms an impenetrable natural fence. Early American settlers created prized bows from the wood.
Red Mulberry	The fruit is similar to a large blackberry in shape, color, and taste. Most creatures find the fruit irresistable.	This tree provides exceptionally delicious fruit; however, humans rarely get to taste it. That is because birds and other animals adore it and will strip the tree of fruit the day before you harvest. If you can bear to chop it down, the tree's wood makes excellent fences.

In a few cases, when confronted with lengthy cells, you can simply edit down the text into more concise sentences. In most cases though, lengthy columns are a signal that you should convert the table into a series of paragraphs. Oftentimes, the leftmost column can serve as a section header for these paragraphs. For example, the following is a prose version of the preceding table.

Osage Orange

Osage orange trees produce a fruit having a bizarre, brainlike exterior. This fruit is about the size of a citrus orange.

The tree has scary-looking spikes. When several trees are planted close together, the hedge forms an impenetrable natural fence. Early American settlers created prized bows from the wood.

Red Mulberry

Red mulberries produce fruit that is similar to a large blackberry in shape, color, and taste. Most creatures find the fruit irresistible.

These trees provide exceptionally delicious fruit; however, humans rarely get to taste it. That is because birds and other animals adore it and will strip the tree of fruit the day before you harvest. If you can bear to chop it down, the tree's wood makes excellent fences.

Rules

A **rule** is a line in a table that separates columns or rows. Table 8-13 contains a full complement of interior and exterior rules.

TABLE 8-13 A Ruled Table

Family	Fruit	Comments
Beech	Small nut	Needs a wet environment
Maple	Double-winged fruit	Grows hard wood
Palm	Gigantic nut	Prefers a warm environment
Pine	Wood cone	Produces turpentine and rosin

If you create ruled tables, follow these guidelines:

- Use very thin lines for rules. Thick rules look clumsy. The only exception is that the top and bottom rules can be thicker to help mark the table's border.

- Provide a slightly thicker rule (or a double line) to separate the header row from the first body row.

The lines on the outside of the table are called *exterior rules*. Those in the middle are called *interior rules*. Some ruled tables omit exterior rules. If your table does contain exterior rules and interior rules, the exterior rules should be the same line width as the interior rules.

Are Ruled Tables Better?

Graphic artists generally detest rules, so rules have disappeared from many technical documents. Ruled tables look cluttered; unruled tables look cleaner. However, horizontal rules help guide a reader's eyes across a row. Horizontal rules (and to some extent, vertical rules) reduce reading mistakes.

Many organizations debate the value of rules. A reasonable compromise is to render tables that contain interior rules but do not contain exterior rules. Some organizations take it one step further, providing only horizontal rules to separate rows and omitting vertical rules that would separate columns.

Are ruled tables better than unruled tables? The answer really comes down to form versus function. Unruled tables provide better form, while ruled tables are slightly more functional.

Shading

Row **shading** provides an aesthetically pleasing and highly effective alternative to rules. For example, consider the shading in Table 8-14.

TABLE 8-14   An Example That Uses Row Shading

Family	Fruit	Comments
Beech	Edible nut	Needs a wet environment
Maple	Double-winged fruit	Grows hard wood
Palm	Nut	Prefers a frost-free environment
Pine	Wood cone	Produces turpentine and rosin

In the preceding table, note the following patterns, which are typical of shaded tables:

- The header row uses a different shading pattern than the body rows. When color is an option, the background color of the header row is usually different from the background of the shaded body rows.

- Every other body row is shaded.

- The bottom and top rows of the table are ruled, which helps visually bound the table.

- Other than the top and bottom rows, shaded tables should not contain rules. Shading eliminates the need for rules.

Captions

Every table needs a table number and a useful caption. After all, readers who are skimming documents are far more likely to read table captions than the table introductions embedded inside paragraphs. Effective captions have the following characteristics:

- They are detailed enough to explain the purpose of the table but are still as concise as possible.

- They avoid repeating all the names of the table's columns (although you often must repeat at least one of the column names).

- They are self-sufficient, not dependent on the surrounding paragraphs. Note that readers' eyes often gravitate straight to tables and skip over the text that introduces the table.

Consider Table 8-15, which describes characteristics of three native North American trees that bear edible fruit.

TABLE 8-15 What's the Best Caption for This Table?

Species	Height of Tree (in Feet)	Fruit
Pawpaw	40	A yellow oblong berry containing several seeds
Black Cherry	50–60	A tiny dark-red drupe, useful in flavoring
American Persimmon	40–45	A soft, pulpy drupe when ripe and a horrid, semi-toxic nightmare when unripe

Suboptimal captions for the preceding table would include the following:

- Native North American Fruit (This caption is not precise; many fruits are inedible.)

- North American Fruit Species, Heights, and Fruits (This caption just repeats the column names.)

A good caption for the preceding table is as follows:

- Characteristics of Some Native North American Fruit Trees

QUANTUM LEAP

Place table captions *above* the table. Place figure captions *below* the figure.

Summary of Tables

When reviewing your tables, ask yourself the following questions:

- Are all tables numbered? The title of each table should contain a number, such as Table 6 or Table 12-5.

- Do all tables have a clear, accurate, helpful caption?

- Does each column contain a column header? Are all the column headers clear?

- Do all numerical column headers contain suitable units of measurement?

- Does the information in the leftmost column provide the best way to organize the table? Is the leftmost column in alphabetical order?

- Is the text within columns grammatically parallel?

- Is the text within certain cells too long? Should these cells be broken out into separate rows? Should the entire table be eliminated and replaced with prose? Is the table itself too long? Should the table be divided into multiple smaller tables?

CHAPTER 9

Graphics

A picture is, of course, worth a thousand words. Unfortunately, creating professional-looking graphics often takes substantially longer than writing a thousand professional-sounding words. Nevertheless, the effort is usually worth it. Effective technical communication engages readers, and nothing engages readers like beautiful photographs, lucid graphs, or exotic illustrations. Few people—if being truly honest with themselves—will deny that they like looking at pictures more than reading text. True, the same people might learn more from the text, but there is something satisfying and easy about pictures, while text requires effort.

FIGURE 9-1 This picture is totally off topic, but you looked at it before you read the opening paragraph, simply because it is a picture.

Time Series

A **time series** is a two-dimensional graph in which the *x*-axis represents the passage of time. The *y*-axis could represent just about anything. For example, consider the time series in Figure 9-2, which plots temperature in Mytown over a 14-day period.

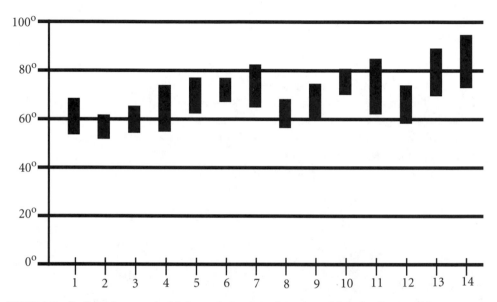

FIGURE 9-2 Daily extremes in Mytown from June 1 to June 14 (a faulty graph).

The preceding time series contains the following faults:

- The *y*-axis spans 100 degrees, yet the actual data only spans about 50 degrees. The compulsion to normalize graphs is very strong; however, starting the *y*-axis at zero wastes space, squishes data, and makes the graph less useful. Since so much space is wasted, it is harder to get an accurate read on individual data points.

- The guiding horizontal lines at 20° intervals are too thick and strong, taking the focus away from where it should be (on the data itself).

- The axes are only partially labeled. Without reading the caption, you can't tell what month this represents.

- The reader gets little sense of how this data relates to the norm. Was this period cooler or warmer than normal?

Figure 9-3 improves upon Figure 9-2.

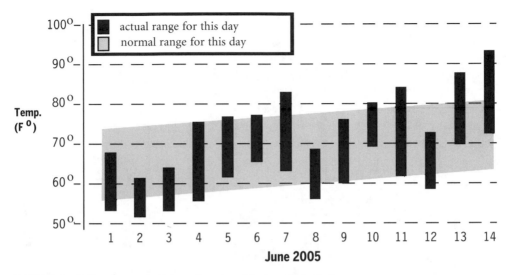

FIGURE 9-3   Daily extremes in Mytown from June 1 to June 14 (better).

Figure 9-3 does a better job of illustrating detail but still does not help readers understand the aggregate. The detail obscures the big picture, making it difficult to determine whether temperatures were warmer or cooler than normal over this period. One simple solution is to add a textual summary announcing the deviation from the norm. An alternate solution is to add a second graph—such as the one shown in Figure 9-4—that focuses on the daily mean. By focusing on the mean, Figure 9-4 makes it easy to see various cool and warm periods and to realize that temperatures were above normal for the period.

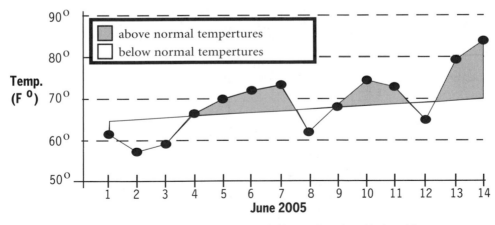

FIGURE 9-4   Actual versus expected mean temperatures in Mytown from June 1 to June 14.

Extra Detail in Online Graphics

The best graphics not only present immediate information but also provide a way for interested readers to dive down to a greater level of detail. Achieving this multilayered presentation is quite difficult in hard-copy graphics but relatively easy in online graphics. In a Web page, as the reader's mouse rolls over a region of the graph, further details about that region can pop up. For example, consider the graph shown in Figure 9-5. As a visitor rolls the cursor over June 6, a second window pops up to display additional information.

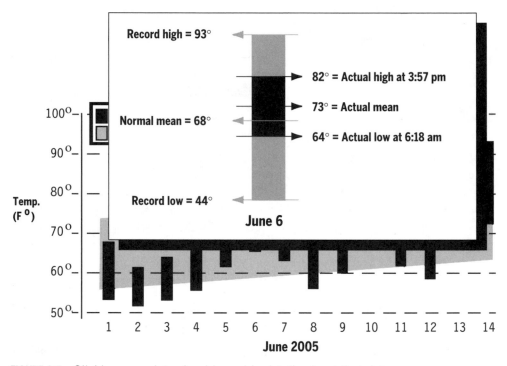

FIGURE 9-5 Clicking on a date should provide details about that date.

It would have been possible for a single graph to represent all the information shown in the primary graph and the popup. However, the resulting graph would have been busy and cluttered. By separating the graphs, each graph stays relatively clean. Casual readers can view what they need, as can more serious viewers.

Before and After

In a before-and-after sequence, lay out the two graphics side by side, with the "before" graphic on the left and the "after" graphic on the right (unless you are writing a manual for an audience that reads text from right to left). For example, consider the before-and-after figure shown in Figure 9-6.

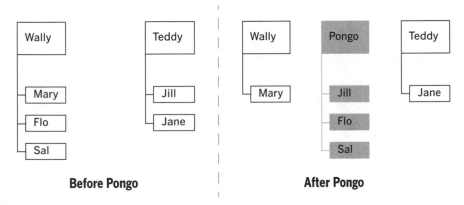

Before Pongo | **After Pongo**

FIGURE 9-6　Walrus mating groups before and after introduction of Pongo.

Many technical communicators mistakenly provide only the "after" picture, leaving readers grasping for the change. The before-and-after sequence in Figure 9-7 is more powerful than either picture by itself.

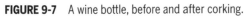

FIGURE 9-7　A wine bottle, before and after corking.

Callouts versus Embedded Text

Labeling the parts of a mechanical or biological unit is typically of great value to readers. To label the parts of a figure or photograph, rely on one of the following two mechanisms:

- callouts

- embedded labels

In a **callout**, the label is some distance away from the part. For example, Figure 9-8 contains four callouts. Most artists connect the label with the part by drawing a line segment. The keys to successful callouts are as follows:

- Keep the line segment short to avoid making the reader's eyes traverse a lot of real estate to connect the label with the part.

- Don't let line segments cross.

- When working in a color medium, render the line segments in a color that contrasts sharply with the figure.

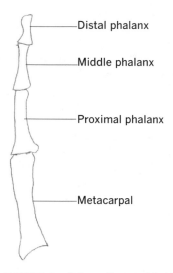

FIGURE 9-8 Using callouts to label bones in the human little finger.

An **embedded label** is affixed to the part itself. In other words, the label is stamped directly on the part. Therefore, an embedded label has no need of a line segment to connect the part to the label. For example, Figure 9-9 contains three embedded labels.

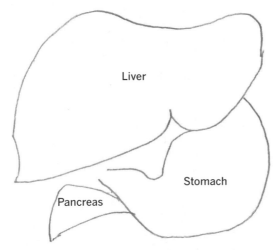

FIGURE 9-9   Using embedded labels to identify parts of the human digestive system.

An effective embedded label should not disturb the reader's view of the part itself. In Figure 9-9, for example, the embedded labels are small enough that they do not interfere with the outline shapes of the digestive organs, although the "Pancreas" label is awfully close to the border.

Callouts versus Embedded Labels

Most professional graphic artists greatly prefer callouts over embedded labels. When the parts you are labelling are skinny, you have no choice but to use callouts. For example, you cannot fit embedded labels inside the bones of your little finger.

As a technical communicator, I'm less concerned about the sanctity of artwork than about communicating effectively. Therefore, I do not feel quite as negatively about embedded labels as most professional graphic artists. In some cases, I feel that callouts force readers' eyes to work unnecessarily hard. For example, when a diagram contains many parts jumbled together like a jigsaw puzzle, it is easier to read embedded labels than to traverse a lengthy line segment to a callout label far, far away.

Graphics That Orient Readers

When walking in a highly touristed area, most people like to see the same map repeated every few blocks. The only thing that changes in these maps is the position of the blessedly comforting words *You are here.* You can use a similar mechanism to orient technical readers. Instead of a map, repeat the same block diagram at the start of each chapter in your book. In place of the magic words *You are here,* help readers get their bearings by highlighting the relevant section of the diagram.

For example, suppose your document contains an introductory chapter followed by three chapters, entitled "Canid," "Coyote," and "Gray Wolf." In this case, consider placing Figure 9-10 on page 1 of the introductory chapter.

Gray Wolf	Coyote
Canid	

FIGURE 9-10   The figure you place in Chapter 1 of the book.

In each chapter, repeat the preceding figure, but highlight the relevant topic. For example, place Figure 9-11 at the beginning of the "Coyote" chapter.

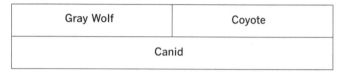

FIGURE 9-11   The figure you place at the beginning of the "Coyote" chapter.

This method of presentation is simple to implement (after all, you only have to draw the picture once) and highly effective.

Screenshots

One way to document a graphical user interface (GUI) is to take **screenshots** of its menus. A screenshot is a "picture" of a portion of the image displayed on a computer monitor. For example, Figure 9-12 shows a screenshot. Most installation and administration manuals feature plenty of screenshots. In addition, online help often contains screenshots.

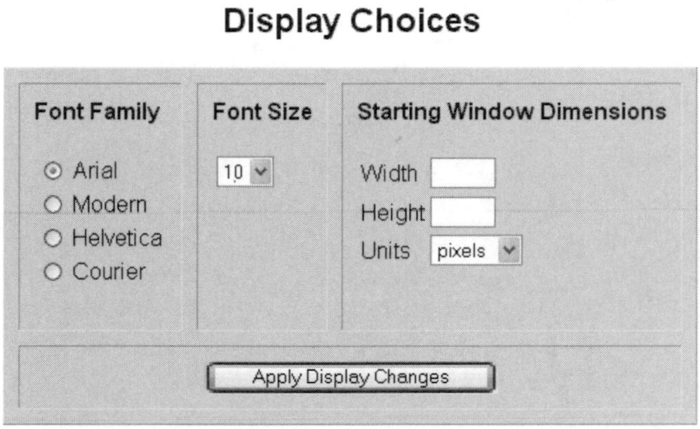

FIGURE 9-12 A sample screenshot.

Writers import screenshots into the documentation. Then, writers provide text explaining what the screenshot illustrates or what the reader should do when confronted with the menu in the screenshot.

Suppose a writer is documenting "Starting Window Dimensions." Good screenshots focus readers' attention on the current topic, so Figure 9-12 is not a great screenshot. After all, it not only shows "Starting Window Dimensions" but also irrelevant, distracting information about "Font Family" and "Font Size." One way to reduce distractions is to eliminate them, leaving only the relevant portion of the screen, as shown in Figure 9-13.

The screenshot in Figure 9-13 is useful only if the reader knows the context ("Display Choices" screen) in which "Starting Window Dimensions" appears. A nice way to have your cake and eat it too is to take a screenshot of the entire screen and then highlight the relevant portion. For example, in Figure 9-14 a thick circle makes it easy for the reader to focus on the "Starting Window Dimensions" section.

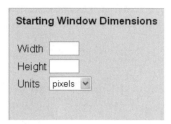

FIGURE 9-13 A subset of the full screen.

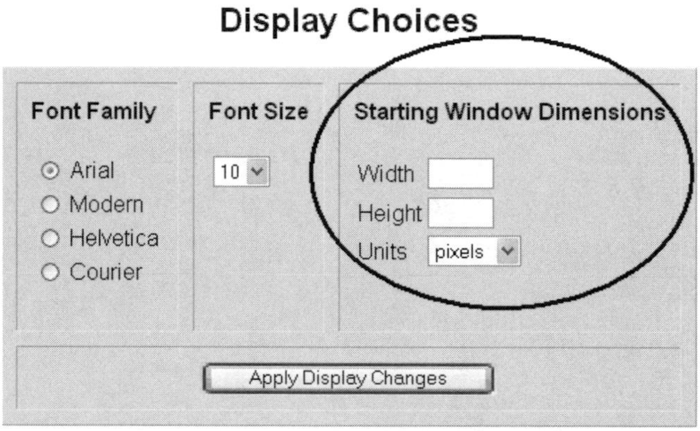

Display Choices

FIGURE 9-14 The full screenshot with the relevant section highlighted.

If you are producing screenshots for a color medium, render the highlighting shape (such as the circle in Figure 9-14) in a color that contrasts sharply with the underlying screenshot. For example, if the screenshot contains a lot of blue and no yellow, then highlighting with yellow will do a great job of catching a reader's eye. If the screenshot is black and white only, then render the highlight shape in red.

The Cost of Screenshots

It is tempting to believe that documentation built around screenshots is cheaper to produce than documentation built primarily around text. In fact, screenshot documentation is more expensive to maintain because screens tend to change more frequently than the accompanying text. Even a tiny change to a screen means that you must reshoot it. (Readers get nervous when a screenshot does not *exactly* match the menu that readers see.) The worst-case scenario is when designers change the background color or background image or font of every screen, which forces the writer to reshoot every screenshot in the book.

Color Blindness

A significant percentage of men with European ancestry are color-blind. A tiny percentage of women with European ancestry are also color-blind, but the condition (like baldness) is primarily a male phenomenon. The wise technical communicator loves the power of color but understands that color-coding frustrates many readers.

Contrary to popular belief, color-blind individuals do not see the world in black and white. Instead, most color-blind individuals perceive green and red as pretty much the same color. All the subtle hues between green and red also blend into a single color.

Nearly everyone—even those categorized as color-blind—can distinguish blue from red. Therefore, if you need to color-code a figure with only two colors, red and blue would be your best choices.

Many maps provide continuous color-coding. For example, consider a map showing the sea surface temperature over a vast ocean. Such maps usually rely on a principle that a leading graphics theoretician[1] calls "the least-perceptible difference." This widely used principle suggests that a map should illustrate a subtle change in value with a subtle (but perceptible) change in color. For example, if a sea surface temperature of 72° is keyed as orange, then a sea surface of temperature of 73° should be keyed as a slightly redder shade of orange. This principle has great aesthetic appeal but is beyond frustrating to the color-blind. In fact, many people without color blindness find it difficult to distinguish close colors.

An alternate approach for continuous maps (and one that will also work in a black-and-white medium such as this book) is to rely on shading instead of color. For example, the continuous map shown in Figure 9-15 shows sea surface temperatures coded with shades of gray.

---

1. Edward Tufte is the world's leading expert on technical graphics. His books are a visual treat.

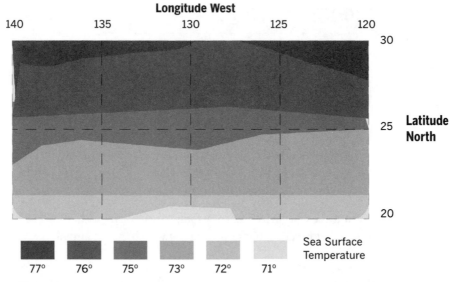

FIGURE 9-15   Sea surface temperatures in gray scale.

Many people find it difficult to distinguish between close shades of gray. Although Figure 9-16 is an aesthetic disaster, most people will find it easier to understand than Figure 9-15.

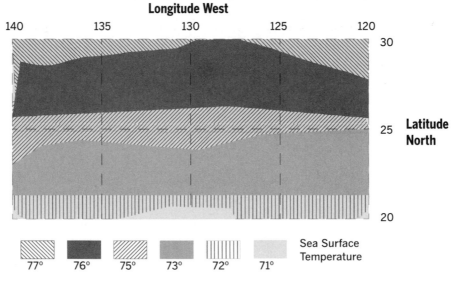

FIGURE 9-16   Sea surface temperatures in gray scale and various backgrounds.

Block Diagrams

Block diagrams typically consist of a set of rectangles (or other geometric shapes) with embedded labels. Artists visually connect the rectangles by stacking them to suggest a hierarchy or by shooting arrows from one rectangle to another.

Block diagrams have become the bread-and-butter illustration of many engineers. Microsoft Visio makes it rather easy to draw block diagrams, but even with a powerful drawing package, many block diagrams still look pretty miserable. The key to creating professional-looking block diagrams is consistency; you have to be consistent both within a single graphic and among all the graphics in a document. To study this principle, consider the rather inconsistent graphic shown in Figure 9-17.

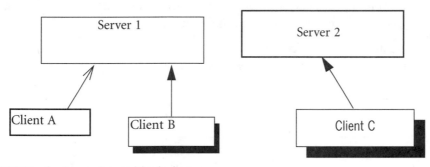

FIGURE 9-17 An inconsistent block diagram.

What's wrong with this picture? Let me list the ways:

- **The text is set in two different fonts.** Notice that Client C is set in a different font than the text in other boxes. It is a good idea to use the same font family within a figure and throughout all the figures in a document. If you need to draw attention to text, then apply italics, bold, or color.

- **The text alignment differs.** Some of the text is left aligned (for example, Client B) but other text (for example, Client C) is centered. Even worse, some of the text is vertically aligned with the top of the box (for example, Server 1) while other text is vertically centered (for example, Server 2).

- **The arrowheads differ.** Needless to say, all the arrowheads in a document should have the same shape, size, and color.

- **The arrows start and end from different sorts of places.** If possible, arrows should all start from and end at the midpoint of a line segment; for example, the arrow connecting Client C and Server 2 connects nicely, but the arrow from Client A to Server 1 is lost in space.

- **Some of the boxes are three-dimensional and others are two-dimensional.** All the elements in graphics should typically have the same number of dimensions. However, if you do mix dimensions, at least render related elements (for example, all client boxes) in the same number of dimensions.

- **The size of the rectangles varies without reason.** There is no compelling reason why Client C is larger than Client A or Client B; all client boxes should be the same size and all server boxes should be the same size.

- **The rectangles for related elements are not aligned.** The three client boxes should align horizontally, as should the two server boxes.

- **The left edge of the graphic is not left aligned.** You should align graphics as closely as possible with the left edge of the body text on the page. In this book, for example, body text is indented one inch from the margin; therefore, the left edge of all figures in this book are indented one inch from the margin.

- **The width of the line segments in the rectangles varies.** Note that the rectangle bounding Client A is thicker than the rectangle bounding Client B.

Figure 9-18 corrects these problems.

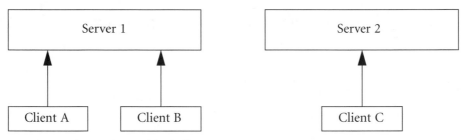

FIGURE 9-18 A consistent block diagram.

The consistency notes mentioned here are mere guidelines. Clever artists can sometimes break these rules and emerge with beautiful figures that still look perfectly professional.

Text That Supplements Figures

The text surrounding a figure should enhance the figure. (For that matter, the figure should enhance the text surrounding it.) This page provides a few principles regarding the prose near a figure.

Keep Illustrations Close to Text

Keep illustrations physically close to the text that introduces them. Don't put the relevant illustration five pages later (except in a lab report where the illustrations, by custom, go at the end).

Focus Attention

The paragraph that precedes a figure should help focus the reader's attention on the figure that will follow. In some cases, the paragraph should highlight a particular portion of the figure that the reader should study. For example, notice how the following text refers to the relevant portion of Figure 9-19.

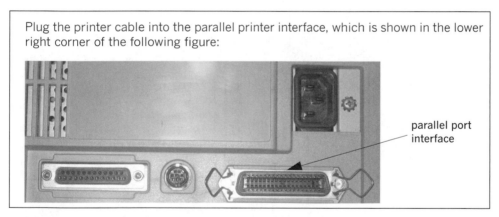

Plug the printer cable into the parallel printer interface, which is shown in the lower right corner of the following figure:

parallel port interface

FIGURE 9-19   The text preceding a figure should introduce the figure.

Use Terminology Consistently

Use the exact same terminology within the figure and the text surrounding it. Failure to adhere to this subtle point has caused readers immeasurable pain. For example, notice the "parallel port interface" callout in Figure 9-19. If the paragraph preceding the figure had referred to this region as the "parallel interface" or the "printer interface," you'd be surprised how many readers would wonder what you were referring to. Even a simple slip like referring to this region as the "Parallel Port Interface" (in uppercase) would trip up quite a few readers.

Technical Photography

Digital photography holds enormous value for technical communication. The following elements have recently come together to make technical photography highly desirable:

- High-resolution cameras and storage media have become relatively inexpensive.

- High-resolution color printing is inexpensive.

- High-bandwidth networks to transfer high-resolution photographs have become ubiquitous.

- Most end users now have programs that display digital photos.

Despite these advances, technical photography remains a challenging medium to master. The primary challenge is figuring out how to use technical photographs to enhance the technical story you are telling. Of course, some photographs are obvious. For example, if you are describing a certain airplane, displaying a photograph of the airplane is always a sound idea. However, suppose you are describing the aerodynamic performance of a new airplane wing. A photograph of the wing alone probably won't be all that valuable.

I would urge you to consult a digital photography manual for details on aperture, shutter speed, and so on. In the next few pages, I only want to focus on a few common techniques for telling a technical story.

Line Art Enhances Technical Photographs

Some people think that photography is photography and line art is line art and the two should never mix. In fact, you should frequently supplement technical photography with line art in order to tell a more effective story.

Please forgive a slight digression on juggling technique. Jugglers do not simply throw objects at random. Instead, jugglers throw objects in a particular sequence, along a particular path, at a particular height, and in a particular rhythm. This entire combination of particulars is called a *juggling pattern*. The juggler shown in Figure 9-20 is performing a juggling pattern called the *five-ball cascade*.

If you study the left side of Figure 9-20, you'll quickly encounter one of the central flaws of relying solely on pure technical photography when capturing dynamic images. The flaw is that it is impossible to tell which of the balls are going up and which are going down. Horizontal motion is also impossible to detect. By adding a little line art (in this case, some arrows), the direction of motion becomes clearer. In fact, the length of the arrows suggests the relative speed of each object. An alternative to line art is to open the camera shutter for a longer period and capture a blur of motion that suggests direction; however, the line-art arrows provide a far clearer way to teach the five-ball cascade.

FIGURE 9-20   Arrows help demystify a complex juggling pattern.

Big Picture First, Then Details

Many movies begin scenes with an **establishing shot** that gives viewers a sense of the big picture. For example, a certain scene begins with an establishing shot of the exterior of a castle. Then the action moves inside the castle where we see actors. The establishing shot provides context; without the establishing shot, viewers might become lost.

Technical documentation that uses photography (or other graphics) should also start with an establishing shot prior to moving down to details. There is a world of wonder in the details, but unless viewers understand the big picture, the details might be confusing. For example, the now familiar Figure 9-21 is a reasonable establishing shot.

FIGURE 9-21   An establishing shot of the whole scene.

One way to dive down to detail is simply to magnify a portion of the big picture. For example, one of the more common questions that juggling students have is, where do you look? Figure 9-22 provides the correct answer.

FIGURE 9-22   Jugglers focus at the top of the juggling pattern.

Of course, don't feel that you can only display magnifications of the establishing shot. You are free to move the camera around, using other angles or showing other parts of the body. For example, another common question from jugglers is, what is the correct stance? Figure 9-23 helps provide the correct answer.

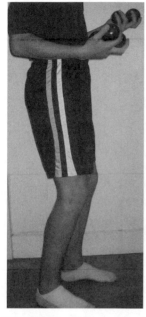

FIGURE 9-23 The proper juggling stance.

Layout: Controlling Focus

Magic is an art form that uses psychological principles to get an audience to look the wrong way. **Layout**[2] is an art form that uses psychological principles to get an audience to look the right way. Learning a few basic layout techniques makes any document or Web site look far more professional. The primary goal of layout is as follows:

> Get your readers to look at the thing you are trying to emphasize.

This chunk explains how to achieve this primary layout goal.

Readers Follow Pointers

When you first look at a graphic, your eyes typically do not focus on random points. Instead, various visual cues guide your eyes toward a focal point. These visual cues are called **pointers**. In illustrations, arrows are pointers. (To a lesser degree, line segments without arrow heads are also pointers.) For example, what do your eyes look at first in Figure 9-24?

FIGURE 9-24   Pointers guide your eyes.

My guess is that you looked at the circle before you looked at either square. (That was the intention of the pointers, at any rate.)

When creating a graphic for paper or a Web site, use pointers to get readers to look at the product or technology you want to emphasize.

Eyes Are Pointers

Humans instinctively look in the direction that other humans are looking. If you suddenly look up, the people around you will also look up. This reaction is so ingrained that humans will even look in the same direction that humans in photographs are looking. In Figure 9-25, for example, your eyes will follow the model's eyes.

---

2. Layout is called *composition* in the fine art world.

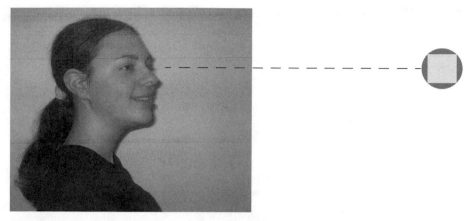

FIGURE 9-25   Your eyes will naturally look at what the model's eyes are looking at.

Readers Are Attracted to Discontinuities

Readers focus on any discontinuities in color, shading, or contrast. For example, in Figure 9-26, your eyes will probably focus on the shaded circle first, even though it is not in the center of the figure.

FIGURE 9-26   Changes in shading also catch your eye.

Humans focus sharply on abrupt changes in color. When working in a color medium such as the Web, render the object you are trying to emphasize in a color that contrasts strongly with the background. For example, if you are trying to emphasize a blue object, put the background in yellow.

An alternate way to use color is to render the entire setting in grayscale except for the object you want to emphasize. Render that object in a sharp color. Your readers' eyes will make a beeline for it.

Layout: Keeping Eyes on the Page

The second goal[3] of layout is as follows:

> Keep your readers' eyes focused on the page.

The preceding goal probably strikes you as a bit odd. After all, if you are reading a book, why would you want to focus your gaze outside the page? For that matter, if you are gazing at a Web page, why would you suddenly turn your attention to another program? Well, let's start with the extreme case shown in Figure 9-27.

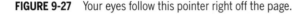

FIGURE 9-27   Your eyes follow this pointer right off the page.

This arrow carries your gaze right off the page. Figure 9-27 would make professional layout artists faint from pain. Remember that human eyes are a powerful pointer; the model's eyes in Figure 9-28 take your own eyes right off the page.

FIGURE 9-28   What is she looking at? You can't tell, but clearly, it isn't on this page.

---

3. Some graphic artists would describe this as the primary goal rather than the secondary goal.

Layout: White Space

White space is the part of a printed page that contains no information. Since paper is usually white, white space literally refers to the white sections of the page, or the sections that do not contain ink. In a Web page, the white space is the background color. Thus, even if the background color is teal, teal blocks of the Web page are still termed *white space*.

A key principle of layout is as follows:

> White space is good.

The corollary to this principle is as follows:

> Clutter is bad.

In a technical manual, readers love white space. Cramming too much visual information into a small space makes readers nervous. White space is an island. White space is a place for the reader to rest before attacking the next topic.

When creating illustrations, always consider white space. For example, consider Figure 9-29, which suffers from a serious white space deficit.

DBMS (Database Management System)	JDBC (Java Database Connector) Driver	J2EE (Java 2 Enterprise Edition) Application Server

FIGURE 9-29 Not enough white space.

Just looking at Figure 9-29 makes me feel claustrophobic. This figure's sins are as follows:

- White space is missing above the top of the figure and below its caption. Notice how scrunched the entire figure appears. Because of the absence of white space, some readers might not even notice the existence of the figure at all.

- The artist has placed too much text inside the three rectangles; the text is cluttered and hard to read. Notice how the text bumps up against the rectangles.

- The rectangles are too close to each other.

Figure 9-30 exaggerates white space somewhat to make a point. Notice that the text inside the rectangles has been stripped down to the essentials. (Text preceding or following this figure should expand the acronyms.)

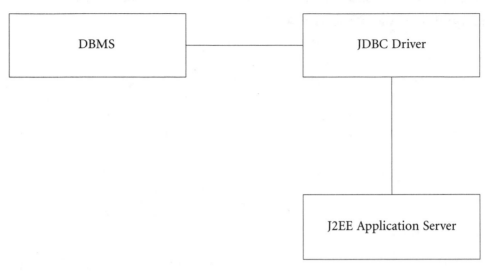

FIGURE 9-30   Rearranging the elements to enhance white space.

Note that it is not always appropriate to recast a horizontal drawing as a vertical drawing because a vertical drawing can imply a hierarchy.

QUANTUM LEAP
Many people center graphics halfway between the left margin and the right margin. However, most graphic artists prefer to lay graphics out flush with the left edge of their captions.

Summary of Graphics

When reviewing graphics in your document or Web site, ask yourself the following questions:

- Does the illustration supplement the body text? Does the body text surrounding the illustration help to explain the illustration? Does the body text use the same terminology as labels inside the illustration?

- Does the illustration contain a caption? By convention, the caption for an illustration is just below the illustration. Does the caption accurately convey the purpose of the illustration?

- If the illustration contains labels, are the labels consistent? For example, a single illustration should not mix callouts and embedded labels. Is the font consistent throughout the illustration?

- Is the illustration working too hard to cram in as much information as possible? An illustration is a little like a paragraph—when it starts to contain too much information, you should divide it in two.

- If this is a graph, are both axes clearly labeled? Are the units clearly labeled?

- Have you optimized the presentation of your graphics by providing a proper layout? Do pointers push readers' eyes off the page? Does the page contain enough white space, or does it look cluttered?

CHAPTER 10

Professional Secrets

This chapter teaches the tricks of the trade. It takes you well beyond the basics and squarely onto the turf of professional technical and science writers. Mastering these techniques will bump your writing up from merely adequate to truly superb.

In this chapter, you will learn the following techniques:

- building explanations for technical readers
- building examples in a variety of ways
- creating documents in question-and-answer format
- describing the same principle in multiple ways
- picking the proper tone
- varying the pace
- creating footnotes and sidebars

What Is Good Writing Anyway?

The best description of good writing that I've ever heard didn't come from a fellow writer but from an experienced engineering manager. He pointed to one of the writers on my staff and said that he "really knew how to tell a good story."

That one line has stuck in my head for many years because it is simple, counter-intuitive, and completely accurate. After all, the trick is not just in explaining something clearly but in doing it so that your audience will actually want to read it and remember it, just like a good story.

Explanations of Formula-Based Rules

Science writers must frequently explain rules or principles, often those based on algebraic formulas. To explain a rule or principle to a *technical* audience, consider using the following algorithm:

1. State the rule or principle formally.

2. Supply the relevant mathematical formula.

3. State the rule or principle casually.

4. Provide an example that shows the rule in action.

5. Describe special cases or limitations with the formula.

6. Provide a succession of examples that use the formula.

To explain a rule or principle to a *lay* audience, focus on steps 3, 4, and possibly 5. Most members of a lay audience will run screaming from anything that involves algebraic formulas.

The following page uses Steps 1 through 5 of the preceding algorithm to explain the rules of gravitational force. The members of the target audience are college freshmen taking an introductory physics course. As always, the wise writer focuses on the interests of the audience; college freshmen might enjoy a hint of romantic attraction thrown in with gravitational attraction.

Note that this is a relatively brief explanation for a topic that requires a much fuller treatment. Following this initial description, a good writer would provide a succession of examples, each one building on the previous one and getting gradually more sophisticated. Most technical readers learn more through good, mathematically based examples than through prose.

Gravitational Force (an example of an explanation)

The **gravitational force** between any two objects is proportional to the product of their two masses and is inversely proportional to the square of the distance between them. The formula for determining the gravitational force between any two objects is as follows:

$$F = (G)(M_1)(M_2)/R^2$$

where:

- F is force, typically expressed in N.
- G is the gravitational constant, which is 6.673×10^{-11} m^3/kg^2s.
- M_1 and M_2 are the mass of two objects, expressed in kg.
- R is the distance between the two objects, expressed in m.

As the formula *indicates*, more massive objects exert a stronger tug on each other than light objects. Thus, two stars pull more strongly at each other than two paperclips. In addition, gravitational force depends heavily on the distance between the two objects. Since this is an inverse-squared proportionality, doubling the distance between two objects will decrease the gravitational force between them by a factor of four.

Gravitational force describes the tug that *any* two objects exert on each other. For example, at this very instant, the stapler and ruler on your desk are lunging towards each other like two tempestuous lovers. Unfortunately for this unlucky pair, their gravitational force is much smaller than the force of friction that holds them glued to their spot on the desk. However, if you were to transport this star-crossed pair into the vacuum of space and place them a few meters apart, they would start moving toward each other, eventually clinging together for eternity.

Example

A photocopier machine having a mass of 125 kg and a laser printer having a mass of 18 kg are 2.5 m apart. What gravitational force do they exert on each other?

```
Given:
  G = 6.673 × 10⁻¹¹ m³/kg²s
  M₁ = 125 kg
  M₂ = 18 kg
  R = 2.5 m


  F = G(M₁)(M₂)/R²
  F = (6.673 × 10⁻¹¹ m³/kg²s)(125 kg)(18 kg)/(2.5 m)²
  F = 2.4 × 10⁻⁸ N
```

When objects are extremely small—down to the subatomic size—the preceding gravitational force formula no longer applies. In fact, a different set of forces takes over.

Examples

Good descriptions require good examples; descriptions without examples are often too abstract for readers. Adding examples renders the abstract more concrete. For example, consider the following wholly accurate (but colorless) description for lay readers:

> The pressure of a gas is proportional to its temperature.

What could be harder to visualize than an invisible gas? When writing for lay readers, it is far better to use examples that describe everyday phenomenon rather than laboratory or mathematical phenomenon. To help readers visualize this principle, expand the description with a familiar example, such as the following:

> The pressure of a gas is proportional to its temperature. For example, have you ever noticed how flat your tires can look on a frosty morning? As the day heats up though, those same tires begin to look robust again, even if you don't pump more air into them. The gas pressure increases because the sun raises the temperature of the air inside the tire.

Did you notice that the example started with a rhetorical question? Rhetorical questions help suck in readers far better than simple statements.

The example has a casual, personalized feel to it. The following phrase is telling:

> ...have *you* ever noticed...

Referring to the reader as *you* suggests that this is a phenomenon within his or her experience, not something obscure or rare.

Examples must be appropriate to the target audience. The preceding example is appropriate for most readers in North America and Europe since such readers usually own some vehicle with tires (bike or car) and experience appreciable diurnal temperature fluctuations. However, this example probably isn't meaningful to readers in equatorial climates since they typically won't experience tire deflations in the morning.

Note that some serious technical journals do not permit the use of *you* in examples.

Examples by Metaphor

Metaphors are not just a literary device—they can also come in handy in technical examples. In a metaphoric example, instead of providing an example of a phenomenon at work, you compare the phenomenon to something analogous. For example, consider the following metaphoric example:

> ## Client-Server Architectures (a metaphoric example)
>
> Many modern computer systems rely on a *client-server* architecture. A client-server architecture consists of the following two kinds of components:
>
> - clients, which are components that make requests
>
> - servers, which are components that fulfill those requests
>
> Since clients and servers are typically on different machines, the requests and responses must flow through a network.
>
> Operations within a typical restaurant are similar to a client-server architecture. The hungry customers are clients. Like all clients, the customers make requests. ("I'd like Pad Thai, please.") The chefs are servers, responding to client requests by preparing sumptuous plates of noodles. Waiters take on the role of the network, shuttling requests and responses between clients and servers.
>
> Client-server architectures bog down when too many clients make requests in a short period. Suddenly, instead of a server responding in two seconds, the server might not respond for ten seconds or more. In fact, the server might be too busy to respond at all. You've probably seen this happen in a busy restaurant. When too many customers sit down around the same time, the wait for food becomes intolerable. At such times, the waiter might even apologize profusely, saying that the chef lost your order.

The preceding metaphor fits like a glove.[1] In fact, the analogy can even apply to special cases, such as the overloaded servers in the final paragraph. The analogous phenomenon (restaurant operations) is familiar to everyone in the audience. Finally, the analogy is kind of fun, since most readers enjoy going to Thai restaurants.

---

1. Well, almost. Since waiters often introduce themselves by saying, "Hi, my name is Fred, and I'll be your server tonight," this analogy could confuse some readers.

Examples for Programming Documentation

Just as many readers skip over text to look at pictures, many readers skip over dry recitations of principles in order to get to examples. This is particularly true when writing documentation about programming. In other words, programmers often gravitate to example programs and only read the prose as a last resort. When writing a programming manual, code the examples first and then write the text around the examples. The keys to producing good examples for programming manuals are as follows:

- Start with a very simple example. Then, gradually layer on complexity in subsequent examples.

- Do not assume that your audience will always read the prose preceding or following the example. Provide meaningful comments within the code itself. Do not, however, overwhelm the code with too many comments; don't comment the obvious.

- Provide crystal-clear variable and function names. For example, naming a variable `NumberOfRecordsProcessed` helps the reader a lot more than naming it `X`.

- Format the code for clarity. Place blank lines before and after discrete blocks of code. Don't forget to indent code properly.

To illustrate these principles, let's consider a sequence of examples for a fictitious programming language named *Fenster*.

Example 1: Hello World

In 1971, Brian Kernighan and Dennis Ritchie, two researchers at Bell Labs, wrote a landmark technical book, *The C Programming Language*. The book opened with a very short example that wrote the phrase *Hello World* to the terminal. Since that time, good programming manuals have begun with an ultrasimple example, often one that writes the phrase *Hello World* to the terminal. Such examples—whether they write *Hello World* or not—have become known as **Hello World examples**.

```
// This example writes the phrase "Hello World." to your terminal.
program HelloWorld  // Every program needs a name

routine start // A routine named start must appear in every program.
begin
    WriteToTerminal("Hello World.\n")  // The \n is a newline character.
end
```

Example 2: Simple Variables

After getting the *Hello World example* to run, readers will want to see a slightly more complex example. For example, the following program creates and uses a simple variable:

```
// This example demonstrates how to declare a variable, assign a value to it,
// and then write its value.
program DeclareAssignAndWriteAnInteger

routine start
begin
  // Declare an integer variable named tacos.
    integer tacos

  // Assign a value to this integer variable.
    tacos = 3

  // Write tacos' value (3) to the terminal. Notice that we don't put double
  // quotes around tacos. If we did, the program would write the word
  // "tacos" instead of the value 3.
    WriteToTerminal(tacos)
    WriteToTerminal('\n')
end
```

Notice the copious comments throughout the preceding example. Like an experienced sergeant leading his troops safely through a mine field, the final block of comments warns readers of a potential pitfall ("... don't put double quotes around tacos.").

Example 3: Input

Since the first two examples showed how to write information, the audience might now want to know how to read information. The third example provides an answer:

```
// This example demonstrates how to read data that the user types.
program ReadAnInteger

routine start
begin
    integer tacos

    WriteToTerminal("How many tacos would you like?") // Prompt the user.
    ReadValueFromKeyboard(tacos)  // Read the value that the user enters.

    WriteToTerminal(tacos + '\n')  // Echo the value that the user entered.
end
```

Question-and-Answer Format Example

Question-and-Answer (Q-and-A) format is exactly what it sounds like—a collection of questions and their answers.

Q-and-A format is a compelling choice when your audience has a scientific bent. Scientists, by their nature, love to solve riddles and to be challenged with tough questions. Scientists will often attempt to answer the question posed in a header, then read the answer to see if they got it right. Technical documentation becomes a sort of subconscious game show. Q-and-A format is also a good choice for a lay audience since many readers find this style a refreshing break from standard prose.

Despite its charm, readers quickly tire of Q-and-A format. Very few long documents are written completely in Q-and-A style; more often, this style is reserved for brief documents or for individual chapters in a lengthy book. Variety generally pleases intelligent readers, so a single Q-and-A chapter in a lengthy book provides a clever change of pace.

The keys to good Q-and-A documents are as follows:

- Imagine that you are writing a play for two characters—one who is slightly naive provides all the questions, while all the answers come from another character who is a bit of a know-it-all. Allow the two characters to take on different personalities. As in a play, try to keep the dialog conversational and somewhat casual.

- As in a play, let the answer to a question flow naturally into the following question. For example, you can end an answer by introducing a new term and then use the next question to ask what the new term means.

- Write casually—don't confuse Q-and-A format with a lab report.

- Keep your answers short. Don't turn your dialog into a monolog. Answers should rarely go over a paragraph and should never exceed a page. If an answer is too lengthy, readers will forget the question. If many of your answers are too long, then consider an approach other than Q-and-A format.

How Does Q-and-A Format Differ from a FAQ?

A FAQ answers frequently asked questions about a certain product or technology. FAQs are similar to a standard Q-and-A document. However, a FAQ is not meant to be read sequentially—it is just a list of questions that readers can access in random order. By contrast, Q-and-A documents are written to be read linearly (from start to finish).

Question-and-Answer Format Example

The following example uses Q-and-A format to explain a firewall product aimed at technophobic lay readers. Most lay readers greatly prefer the Q-and-A format to a more traditional technical manual. The style is highly conversational and easy to read. Note that this tone is generally inappropriate for a serious technical audience.

InfiGuard (an example of Q-and-A format)

This document explains how the InfiGuard Firewall protects your PC.

Why do I need a firewall?

A firewall protects you from individuals trying to harm your PC or steal data from it.

But I'm a good person. Why would someone want to damage my PC?

We know you're a good person. We only sell to good people. However, there are some very bad people on the Internet.

What could they do to my PC?

Unspeakable things. They could render your PC completely unusable. Or, they could mess up your PC so badly, you'd wish it were dead. Or, they could spy on you and take note whenever you went to a questionable Web site.

How does a firewall protect my PC?

A firewall only lets trusted people and organizations gain access to your PC (and only to certain benign portions of your PC, at that). The firewall prevents any untrustworthy people and organizations from gaining access to your PC.

How does the firewall know who to trust?

The firewall will ask you.

Do I need a degree in computer science to answer its questions?

No. The firewall asks you in plain English. For example, the firewall might ask you a question such as, "Dexco Unlimited is trying to install a file on your system. It looks suspicious. Should I prevent Dexco Unlimited from installing the file? (Yes or No.)"

Is it really that easy?

About 95% of the questions really are just that easy. A few questions are more complex, but the InfiGuard Firewall will recommend an answer for the really tough ones.

In Other Words

We've all had bad teachers who could only explain things in one highly specific way, almost as if they were teaching the class from a script. For example, consider the following classroom dialog:

> Teacher: Cells contain units called organelles.
>
> Student: I don't understand.
>
> Teacher: Cells contain units called organelles.
>
> Student: [Realizes teacher is worthless and that further questions would be pointless.] Yes, I see.

The best teachers know that different students latch on to different kinds of explanations; therefore, great teachers explain each key concept in a variety of ways. Unfortunately, many writers explain key concepts in only one way. With only a single explanation, befuddled readers are forced to reread the same passage, which doesn't get them anywhere.

The phrase *in other words* is a valuable tool. Although the first sentence in the following example is sufficient for many readers, the second sentence should handle any stragglers:

> Hydras reproduce asexually. **In other words,** a single hydra can reproduce all by itself; it does not require a partner.

Although *in other words* is a handy phrase, not all second-chance explanations require it. The phrase *that is* can also be quite helpful, as in the following example:

> Sexual reproduction generates more diverse offspring than asexual reproduction; **that is**, offspring of sexual reproduction differ from their parents, while offspring of asexual reproduction are genetically identical to their single parent.

Tone

Tone is to writing what mood is to a social gathering. Some social gatherings are serious, unemotional, and businesslike, while others are light and filled with laughter. Similarly, although the tone of most technical documents is serious, there are times when readers would prefer a lighter tone. The trick is to assess your audience's mood and figure out what kind of party to throw.

For example, consider the following troubleshooting instructions from a consumer digital camera manual written with a serious tone:

> The most common cause of perceived failure in the Carambola 5000 digital camera is an uncharged battery. Batteries not only discharge rather rapidly with use but also discharge during periods of disuse.

The preceding passage features good, short, active-voice sentences; however, the formal tone is off-putting. It is as if the writer showed up at mardi-gras wearing a three-piece suit. Remember—this is a manual for consumers taking pictures of their pet poodle, not a manual for cardiologists learning to perform bypass surgery. Flipping the passage to a lighter tone yields the following:

> We at Carambola Industries are concerned with our self-esteem and want to do everything possible to prevent the following dialog:
>
> You: My Carambola 5000 is broken!
>
> Us: [Concerned.] Is the battery charged?
>
> You: It must be. I charged it up and haven't used the camera for six months.
>
> Us: If the camera hasn't been used for six months, then the battery isn't charged. That battery can only hold a charge for about three months.
>
> You: So, the camera isn't broken?
>
> Us: [Wiping tears away.] No sir. It is likely in fine order.

Most people only learn to write in one tone. Experienced writers master multiple tones and shift among them as required. Although a single document is generally written in a single tone, you may need to shift tones periodically. For example, even a light-toned consumer manual should turn deadly serious when discussing safety issues.

Pace

Pace is the speed at which a document presents information. Given a certain set of facts to present, a fast-paced manual covers the material in fewer pages than a slow-paced manual. Pace is information density.

What pace is best? Mastering pace in writing is a little like mastering a car-racing video game. The trick is in knowing when to slow down and when to speed up. Generally speaking, intelligent readers prefer a fast pace. This book, for example, is written at a rather fast pace. However, even highly intelligent readers prefer a slow pace when learning something highly complex or confusing. Thus, this book slows down occasionally to bathe in the tougher concepts. To slow things down, consider explaining the same concept multiple ways or providing multiple examples.

Many writers change pace just for the sake of change. If done well, changing pace helps engage your audience. The following techniques are helpful for varying pace:

- Mix in a short paragraph between a couple of long ones. Keeping each paragraph the same length is like delivering a speech in a monotone—you'll just put your audience to sleep.

- Break up long strings of paragraphs with some bulleted and numbered lists.

- Sprinkle tables throughout a document; don't clump them.

- Break up long blocks of text with an occasional figure or photo.

- Use sidebars.

What Are Sidebars?

These shaded blocks of text sprinkled throughout the book are called **sidebars**. They are common in newspapers and magazines but somewhat rarer in technical documentation. That's too bad—readers' eyes feel an almost gravitational pull toward the distinct shading in sidebars.

Authors use sidebars for interesting digressions that would otherwise become non sequiturs if placed in a traditional paragraph. Some authors also use them just to make sure they have your attention when making an important point.

Footnotes and Other Digressions

Sigmund Freud,[2] a prolific and superb science writer who had a gift for engaging audiences with fascinating examples,[3] loved footnotes. **Footnotes** are, of course, those tiny numbered notes down at the bottom (the foot) of the page. Freud loved them so much that some pages actually contained more text in footnotes than in standard paragraphs. He can be forgiven since footnotes were all the rage in scholarly works about a century ago. Plus, he was a writer easily whisked away into digressions, intriguing anecdotes, and linguistic escapades,[4] all of which could find a satisfying release in footnotes.

Modern writers use considerably fewer footnotes than those of Freud's time. In fact, many editors have banished them from documents altogether. You rarely see them in undergraduate textbooks and only occasionally see them in commercial or corporate manuals.[5] However, some academic journals still require citation footnotes or endnotes.[6]

Many modern readers associate footnotes with "boring academic stuff" rather than with the "really cool stuff" for which Freud generally used them. So, what happened to all the really cool stuff? In fact, a lot of it disappeared—a victim of the incessant need to present and ingest information as rapidly as possible. The remaining really cool stuff has crawled up from footnotes into sidebars.

Since this book has hounded you to be concise, you might wonder if there is still a place for anecdotes and curious digressions, or whether such information should stay repressed. By way of explanation, suppose you must travel from Florin to Guilder.[7]

FIGURE 10-1   What is the best route from Florin to Guilder?

---

2. Freud, the founder of psychoanalysis, lived from 1856 to 1939.

3. Of course, it is pretty easy to fascinate an audience when you are writing about sex.

4. He spoke eight languages and could always come up with the perfect adage in one of them.

5. This manual, for example, uses noncitation footnotes, primarily as an outlet for weak jokes.

6. **Endnotes** are just like footnotes except that they appear in a clump at the end of a document rather than at the bottom of the source page.

7. You might recognize these as the two featured countries in William Goldman's novel *The Princess Bride*. Goldman took these pretty names from the two terms for pre-Euro Dutch currency.

The map in Figure 10-1 suggests that you can either take the direct (concise) route or the scenic (but digressive) route. The wise writer allows the reader to take the direct route when hurried and the scenic route when time permits. From a pedagogical perspective, the scenic route is always more memorable.

Beyond the Obvious

Good technical writing doesn't take the easy way out by just stating the obvious; good technical writing always seeks to tell the whole story. For example, consider the following description from a photo-editing software manual:

Edit -> Convert -> ColorToGrayScale

The ColorToGrayScale menu option converts a color image to grayscale.

Note that the target audience for the preceding passage consists of professional and semi-pro photographers who are already quite knowledgeable about many aspects of digital photography. The preceding description is accurate, but it doesn't really tell these readers anything that they don't already know. Compare the previous description to the following:

Edit -> Convert -> ColorToGrayScale

The ColorToGrayScale menu option converts a color image to grayscale, More precisely, the function maps each screen pixel from its current RGB value to a value in which R, G, and B are all identical, which is always a shade of gray. The resulting black-and-white photo can display up to 256 different shades of gray ranging from pure black to pure white to everything in between.

The conversion algorithm relies on the *intensity* of each color pixel. For example, a bright blue ball and a bright red ball of equal intensity will look identical when converted using this algorithm.

The conversion algorithm is not the same one that most digital cameras use when taking a picture in black and white mode. Our conversion algorithm tends to produce slightly brighter (more white) images than most digital cameras shooting in black and white mode. We have found that our algorithm tends to look better on screen, and the digital camera algorithm tends to look better when printed.

As you can see, the fuller explanation attempts to answer the questions that the target audience would have. The earlier one-line explanation is worthless.

Even the fuller explanation could go a lot further. For example, a more complete explanation would show different sample images in color and then converted to black and white, along with explanations.

Precision Descriptions

Many of the comic books that I read as a kid contained an advertisement for a correspondence art school. The advertisement showed a picture of a deer. His name was Winky. If you wanted to be an artist, you drew a picture of Winky and mailed it to the admissions committee. If you could draw Winky accurately, then you had the right stuff to become a famous comic book artist.

I believe that ads for a correspondance technical writing school should also feature Winky. Except, instead of drawing Winky, you would describe him in words. If you could paint a verbal picture so perfectly that someone who had never seen a cute wittle deer could picture Winky exactwy, then you would be granted admission. After all, good technical and scientific writing often requires precise verbal descriptions.

Do you have what it takes to enroll in the Famous Technical Writer Correspondence School? If you feel good about your chances, take out a sharpened QWERTY keyboard and describe exactly what you see in Figure 10-2. Your explanation should be so good that someone with a scientific or technical background could understand exactly what you are looking at. (No fair looking at the next page until you have completed your test.)

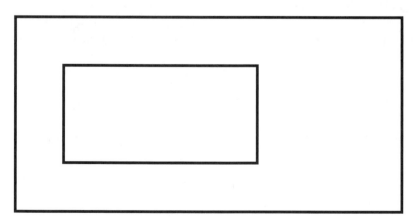

FIGURE 10-2   The admissions test.

The following is a possible description of Figure 10-2:

> The figure contains two rectangles, referred to as follows:
>
> - the outer rectangle
>
> - the inner rectangle
>
> The outer rectangle is 4 inches wide by 2 inches tall; the inner rectangle is 2 inches wide by 1 inch tall.
>
> The inner rectangle is completely inside the outer rectangle. The top and bottom sides of the inner rectangle are parallel to the top and bottom sides of the outer rectangle. The left edge of the inner rectangle lies 0.5 inches to the right of the left edge of the outer rectangle. The top edge of the inner rectangle lies 0.5 inches below the top edge of the outer rectangle.
>
> The perimeters of both rectangles consist of pure black lines that are $\frac{1}{36}$ inch thick. Other than the rectangles, everything else in the figure is pure white.

Notice that the preceding description begins by labelling the two parts and sticking to these same part names throughout the description.

You would probably hate to read an entire book written with so little imagination or verve; however, in some situations, a completely objective and robotlike description is exactly what scientific readers need.

The Hardest Part of Writing

For many people, the toughest part of any writing project is the beginning. This is only natural. At the beginning of a writing project, a desolate white screen is staring back at you. You have not yet established a track record with this document. Doubt gobbles at your soul. "Hmm, maybe I'll take a break and get a snack now. Oh look—celebrity bowling is on Channel 4!"

Many people make the mistake of trying to write books linearly, from page 1 to page *n*. Don't do this. Professional writers almost never do this. Page 1 is almost always the hardest page to write. In fact, sentence 1 of page 1 is the hardest line to write. Why put so much pressure on yourself? Write the first chapter last. By the end of the project, the opening chapter will fly off your fingertips.

What should you write first? I recommend writing the easiest chapter or section first. Build some confidence in yourself before tackling the hardest parts.

Writer's Block

Have you ever had writer's block? Of course you have. Everyone has. Even Stephen King has days where he can only write a single novel.

Try the following tactics to overcome occasional bouts of writer's block:

- Try writing passages without self-editing. In other words, just try blasting out a bunch of sentences and force yourself not to erase or rewrite anything during this blast.

- If you've gone more than 15 minutes without writing anything, step away from the computer and leave the typing room for half an hour. Take a shower, take a walk, talk to a friend... anything! Just make sure to engage your brain with something other than writing for 30 minutes. Then come back and try again.

- Reread something you wrote that you are proud of. Remind yourself that you've succeeded before as a writer and that you will, therefore, succeed again. Writer's block smacks of self-doubt. By removing the self-doubt, you can often accomplish great things.

- Figure out what times of day you are most productive as a writer and restrict your writing to those times.

The most general advice when faced with writer's block is simply to change your process and get out of the rut you're in.

Summary of Professional Secrets

When reviewing your professional documentation, ask yourself the following questions:

- Have you used enough examples? (It would be hard to have too many examples.)

- Have you used appropriate examples for your audience? Will your audience struggle trying to understand your point? Conversely, are your examples too trivial and obvious for the target audience?

- Have you presented examples in the right sequence? For instance, does the first example require information that can only be learned by reading the second example?

- If you have used Q-and-A format, do your questions cover the topic sufficiently?

- Have you only described important concepts in a single way? Hopefully, you have described key concepts in terminology in multiple ways so that readers who don't understand the first explanation might understand the second explanation.

- Have you picked an appropriate tone for your documentation? Are you serious when your readers want something light? Do you put out whoopie cushions when your audience is in no mood for levity?

- Have you picked an appropriate pace? Do you write too quickly or too slowly? Do you slow the pace to handle tricky concepts?

- Assuming that the pace and tone permit, have you provided sidebars, footnotes, or other digressions to wake your readers up? Will your digressions interest your audience or are they just eating up paper?

- Have you written obvious information that your target audience already understands or do you take a step beyond what they already know? Is your documentation worth their time?

- Have you accurately described that which requires objective description?

Writing: Specific Kinds of Documents

This section explains how to produce specific kinds of documents that the average scientist or engineer would have to write, including the following:

- manuals
- Web sites
- proposals
- specs
- lab reports
- PowerPoint presentations
- e-mail messages

 CHAPTER 11

Manuals

A **manual** is a document that explains how to perform a task, use a product, or master a technology. A document detailing the history of a certain car model *is not* a manual, but a document describing how to drive that car or change its oil *is* a manual.

Good manuals are priceless; bad manuals are torture. As a technical person, you already have valuable experience with both kinds of manuals. When writing a manual, emulate what you like, and avoid what you detest.

The following is a list of the most popular styles of manuals:

- cookbooks
- tutorials
- guides
- references
- nonverbal manuals

No style is inherently better than another, although certain styles are better in certain situations. Experienced writers master several styles and choose the appropriate one for each situation. A thorough documentation set includes several of these styles. Alternatively, a single manual can sometimes include several styles. This chapter explains them all. In addition, this chapter examines online help and release notes, which are two types of documents related to manuals.

This chapter concludes by describing the following common components of manuals:

- prefaces
- glossaries
- tables of contents
- indexes

Manual Style: Cookbooks

A culinary cookbook consists of a set of recipes, where each recipe contains the following two sections:

- a bulleted list of ingredients
- a numbered list of steps

Cookbook-style manuals also provide a bulleted list of ingredients and a numbered list of steps. Cookbook-style manuals are ideal for describing tasks that the reader must perform in a highly specific way. For example, a manual explaining how to create a certain chemical compound from two other compounds lends itself perfectly to the cookbook style. The manual would specify the list of ingredients (source compounds, goggles, Bunsen burners, thermometers, and so on) and the steps to synthesize the new compound.

Avoid cookbook style when the technology cannot be described as a series of discrete steps. For example, cookbook style is not appropriate for teaching someone how to code in a computer language.

The keys to writing good cookbook-style manuals are as follows:

- Think through the list of ingredients carefully. Writers are usually pretty good at remembering materials but not so good at remembering that equipment and energy are also some of the ingredients.
- Keep each step discrete. Do not cram multiple tasks into a single step. See Chapter 7 for more suggestions about creating good lists.
- Define success, as in the following example:

If the cells are glowing with a distinctly greenish tint, then you have performed steps 1–24 correctly and can proceed to the next section.

- Define failure and explain what to do about it, as in the following example:

If the cells are not glowing green, then you must perform steps 19–24 again, paying particular attention to the temperature of the enzyme in step 19.

The following is a cookbook-style software manual for experienced users. Notice the descriptions of failures (in steps 4 and 7) and the description of success (in step 7).

Cookbook Example: Installing the Carambola Server

Requirements

To perform the installation, you need a PC with the following minimum requirements:

- CPU— Pentium at 1.5 GHz

- RAM—512 MB

- Network—10 MB/s

- Operating system—Red Hat AS3 or later

- Browser—Mozilla 1.7 or later

Installation

To install the Carambola Server, perform the following steps:

1. Start Mozilla.

2. Browse to the following URL:

   ```
   http://www.carambolapublishing.org/ws/download
   ```

   The Download menu appears.

3. In the Download menu, click the Download and Save menu option.

   The server downloads the software. If the download takes longer than 20 seconds, then a network failure has doomed the download. Wait a few minutes, then start over at step 1.

4. Start a bash shell.

5. Invoke the following command from the bash shell:

   ```
   $ /tmp/inst_carambola
   ```

   The inst_carambola script will install and start the Carambola Server, overwriting any previously existing installation of the Carambola Server.

6. In Mozilla, browse to the following URL:

   ```
   http://127.0.0.1:1984
   ```

   If Mozilla displays an orange fruit, then you have successfully installed the Carambola Server. If Mozilla does not display an orange fruit, terminate the bash shell and redo steps 4 and 5.

Manual Style: Tutorials

A **tutorial** is a document that takes readers by the hand and leads them gently through the initial use of a product or technology. The keys to writing good tutorials are as follows:

- Give readers a taste of success as soon as possible, preferably on the first page. Reassure readers that they can learn this technology.

- Take very little for granted. Do not leave out information that a beginning reader would find valuable just because it seems obvious to you.

- Omit extraneous background information. Readers don't need to understand geology to learn how to dig a hole.

- Move fairly slowly, particularly in the beginning of the document. Readers who get lost on page 1 will not make it to page 2.

- Call out everything that is confusing or potentially dangerous.

- Test your tutorials on beginners; feedback from experts is worthless.

- Introduce topics one at a time, if possible. Do not introduce multiple topics in the same lesson.

If you are an expert who is writing a tutorial, you must regain the beginner's mind. You must regain your initial sense of confusion and imbalance with this technology. It is sometimes easier for a beginner to write a tutorial than for an expert.

Although tutorials are aimed at readers who are new to a technology, your audience might already be expert in an allied technology. For example, when writing a tutorial for a new surgical device, the surgeons in your audience are already highly experienced with related devices. Therefore, in your tutorials, take advantage of existing experience by drawing comparisons to familiar technologies. Doing so will build a sense of comfort with the new technology.

Remember that tutorials are often the first impression (that critical first impression) that readers will have with a technology. A good tutorial can imbue readers with a sense of competence. However, some readers find tutorials intimidating. (*If I don't understand the tutorial, I'll feel really stupid.*)

The following tutorial is aimed at readers who are fairly comfortable using PCs (for example, they already understand the concept of pathnames), but who have no experience creating Web pages with HTML.

Tutorial Example: Getting Started with HTML

Most people create Web pages in a language called HTML. Don't worry—HTML is not a full-fledged computer language; you do not need a degree in computer science to use it. In fact, creating Web pages with HTML can be a lot of fun.

In this first lesson, you will create a Web page. To perform this lesson, you'll need a personal computer (PC) running Windows. The version of Windows you use isn't important. If your PC is newer than 1995 or so, you are all set. Your PC must have a browser installed. A browser is a program that allows you to access Web sites. The most popular browser is named *Internet Explorer*, and it is installed on almost every modern PC. However, your PC might also have a different browser (such as Netscape or Firefox). For this lesson, any browser is fine.

To begin the lesson, you must invoke the text editor named *Notepad* by selecting `Start->Program->Accessories->Notepad`. The Notepad program will spring to life. Enter the following text into Notepad:

```
<p>HTML is better than ice cream.</p>
```

Enter that text *exactly* as it appears. Notice that the closing `</p>` contains a slash, but the opening `<p>` does not.

Congratulations—you just created the HTML for a Web page. To save this HTML, select `File->Save`. A menu appears, asking you to select the file in which to save this masterpiece. Save the file at the following pathname:

```
C:\Temp\IceCream.htm
```

Note that the filename *must* have the suffix `.htm`. If you give the file some other suffix (or omit a suffix), then this lesson will not work.

You are finished with Notepad, so feel free to close it.

To end this lesson, open your Web page in a browser. There are several ways to do this. One fairly simple way is to do double-click `My Computer`, and then to navigate to the `C:\Temp` folder. After arriving, double-click on the `IceCream.htm` document. If you did everything properly, your browser should display a Web page containing the following text:

```
HTML is better than ice cream.
```

Notice that your browser does not display the `<p>` or the `</p>`.

Manual Style: Guides

A **guide** is a manual that readers turn to after reading a tutorial in order to master intermediate or advanced topics. Sometimes a tutorial and guide are bundled together within the same manual. In such cases, the tutorial is generally the opening chapter and the guide contains all subsequent chapters.

Guides and tutorials are useful for products where the steps aren't obvious or where each reader could use the technology in a different way. The keys to writing good guides are as follows:

- Empathize with readers. Determine what skills your audience wants to learn and provide them.

- Provide plenty of relevant examples. Provide a large collection of short, simple examples rather than a small collection of lengthy, complex examples. However, sprinkling in an *occasional* lengthy real-world example is a good idea. Pick examples that will make sense to your target audience.

- Include plenty of figures, tables, and photographs to break up the monotony of large blocks of text.

- Provide background information when it helps explain a topic.

Structured documentation is a popular style of guides. The central organizing unit in structured documentation is a **chunk**, which is a short, discrete unit or lesson on a particular topic. When creating structured documentation, the writer begins by organizing all material into chunks. Then the writer follows these rules:

- Each chunk builds on the previous chunk.

- Each chapter consists of a series of related chunks.

- Chunks usually begin at the top of a new page.

This guide (and all other guides in the Spring Into... Series) are written as structured documentation in which each chunk is either one or two pages long.

The following example demonstrates a single chunk of a structured documentation guide on HTML. It is aimed at a moderately experienced HTML developer. Assume that the chunks preceding this chunk have explained the <tr> and <td> tags.

Guide Example: Creating HTML Headers

Each cell in the top row of a table should be a header. In some tables, the leftmost column of every row should also be a header. To create a header in an HTML table, specify the <th> (short for table header) tag.

You can place a <th> tag anywhere you would put a <td> tag. The <th> tag behaves just like a <td> tag, except the resulting text is usually centered and in boldface. Typically, you place <th> tags in the first row of a table. You might also put <th> tags in the first column of each row.

For example, Example 11-1 uses the <th> tag to create headers for both columns of a two-column table.

EXAMPLE 11-1 The <th> tags specify *Movie* and *Description* as headers

```
<table border=1>
    <tr>
        <th>Movie</th>
        <th>Description</th>
    </tr>
    <tr>
        <td>Casablanca</td>
        <td>Bogie must choose between Bergman and the right thing
            to do.</td>
    </tr>
    <tr>
        <td>Wizard of Oz</td>
        <td>A depression-era Kansas teenager chooses between ultimate
            power and the right thing to do.</td>
    </tr>
</table>
```

The resulting HTML table looks as follows:

Movie	Description
Casablanca	Bogie must choose between Bergman and the right thing to do.
Wizard of Oz	A depression-era Kansas teenager chooses between ultimate power and the right thing to do.

FIGURE 11-1 An HTML table containing proper headers.

Manual Style: Reference Manuals

Like structured documentation, **reference manuals** organize material into a set of focused topics. The topics in a reference manual are typically organized in alphabetical order, much like a dictionary or encyclopedia, although some reference manuals arrange reference pages hierarchically. Each topic in a reference manual describes a discrete part or component of a system. For example, consider an engine that contains 100 parts. The reference manual for this engine would contain 100 reference pages, where each reference page detailed a single part. By contrast, the guide for this engine would likely focus on how to maintain the engine as a whole or how to maintain different engine subsystems.

The keys to writing good reference manuals are as follows:

- Use the same format for every reference page. In other words, each reference page should contain the same set of headers.

- Provide examples on each reference page, if possible. Many writers shy away from examples on reference pages; however, most readers love examples.

- Provide abundant hyperlinks or cross-references pointing to related topics.

- Provide diagnostic information. What error codes are emitted when this part fails?

- Think of reference pages as reminders, not tutorials.

Reference manuals are quite popular for describing software such as Application Programming Interfaces (APIs) and shell-level commands. Typically, software reference manuals are distributed online only.

Programs now generate many reference pages automatically. For example, a program called javadoc automatically generates reference manuals for Java APIs by examining Java source code and programmers' comments. Using a program like javadoc saves a tremendous amount of time and synchronizes changes to software with changes to documentation. In addition, these programs do a better job of capturing certain kinds of details, such as the signatures associated with APIs. However, the resulting reference manuals are only as good as the comments that programmers have provided. Niceties such as diagnostics and example code seldom make it into automatically generated reference manuals. In addition, many software engineers do not put enough time into making the comments clear.

The following example illustrates a reference page aimed at mathematicians who are comfortable with the UNIX operating system.

Reference Example: The `pr1me` Utility

Name
pr1me—determines whether a given number or range of numbers is prime.

Synopsis
pr1me [-v] *n1* [*n2*]

Options
-v
Verbose mode. Displays output as a full word (for example, "PRIME").

Description
Use pr1me to test one or two integers for primality. If you specify one integer, pr1me tests just that number. If you specify two integers, pr1me tests all the numbers in that range. By default, pr1me returns P if the specified number is prime and C if the specified number is composite.

Troubleshooting
Values of n1 and n2 must be integers between 2 and $(2^{64} - 1)$, inclusive. If you specify a number outside this range or if you specify a noninteger, pr1me returns X (or "OUTSIDE RANGE" in verbose mode).

Examples
The following example uses pr1me to indicate that the number 11 is prime:

```
$ pr1me 11
P
```

The following example tests the integers between 11 and 13, inclusive:

```
$ pr1me 11 13
P
C
P
```

The following example generates verbose output:

```
$ pr1me -v 11
PRIME
```

See Also
lcd, gcf

Manual Style: Nonverbal Manuals

As its name implies, a **nonverbal manual** contains no words; it contains only illustrations and photographs. Creating a professional nonverbal manual generally requires great artistic skill. Nevertheless, any of the following situations lend themselves to nonverbal manuals:

- A large percentage of the target audience is illiterate.
- The cost of translating text is prohibitive.
- The skill to be mastered is best described through figures or photographs. (Frankly, words would just get in the way.)

At the risk of stating the obvious, I should say that nonverbal manuals are only appropriate for products or technologies that can be described pictorially. For example, a computer language manual is hardly a good candidate for a nonverbal treatment. However, physical activities such as the following would be good candidates:

- simple medical procedures, such as certain kinds of first aid
- assembly manuals, such as a manual showing how to assemble a bicycle
- various physical activities, such as how to use exercise equipment

Generally speaking, nonverbal manuals should be quite short, as in Figure 11-2 that shows how to put on a piece of safety equipment.

  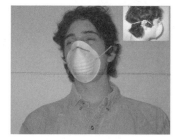

FIGURE 11-2   A nonverbal manual showing how to put on a face mask.

Online Help: Overview

Many readers prefer **online help** to traditional manuals, primarily because good online help brings practical answers quicker than traditional manuals. Good online help is like going to an information booth and getting an instant, if terse, answer from someone who knows it all but doesn't feel obligated to tell you all. To many modern readers, traditional manuals are like going to your Uncle Charley who forces you to listen to the entire history of suspenders when all you want to know is how to make another notch in your belt.

The Differences between Online Help and Traditional Manuals

It used to be easy to distinguish online help from traditional manuals because traditional manuals were distributed in hard copy. Now, however, the line between online help and traditional manuals is blurry. Most product manuals now ship in soft copy, often in an HTML format that looks rather similar to online help. The few remaining distinctions between online help and traditional manuals are as follows:

- Online help pops up when a user selects a Help menu; traditional manuals typically do not.

- Online help does not contain page numbers; traditional manuals do.

- Online help is authored with special tools (for example, Macromedia RoboHelp); traditional manuals are usually written with word processors such as Microsoft Word.

Categories

Online help falls into two broad categories:

- **Context sensitive.** These automatically display a help screen relevant to the reader's current situation. For example, suppose you are looking at a menu to set screen properties. If you request help, a context-sensitive help system displays a help file about setting screen properties.

- **General.** These provide multiple help files and require the reader to pick the right one. If you are looking at the menu to set screen properties and request help, a general help system offers you a list of all help files and you choose between them.

Context-sensitive help is obviously more valuable to readers than general online help. As you might expect though, it is much harder to produce context-sensitive help than general help. Creating context-sensitive help requires careful coordination of writing and software engineering. Of course, if the software developer is also the writer...

Online Help: Best Practices

I recommend the following approach for creating context-sensitive or general online help:

1. Create a list of all the tasks that readers might need to perform.

2. Write a separate help file for each task. For example, if your list consists of 100 tasks, then create 100 separate help files.

3. Add hyperlinks to lead readers to related help files.

4. Beta-test and refine as necessary.

We now dive a little deeper into the preceding list.

Create a List of Tasks

The engineering team should sit down together and figure out what users will do with the product. When creating the list of tasks, keep each task highly discrete. Don't assume that a single menu entry in a graphical user interface corresponds to a single task—it may correspond to multiple tasks (or to none at all). In addition to tasks, you might also want to create a list of concepts. If your list of concepts is long, put the concepts into a traditional manual instead.

Write Help Files

The keys to writing good help files are as follows:

- Keep each help file as brief as possible. Ideally, each help file should fit inside a single screen so that readers don't have to scroll.

- Prefer numbered lists to bulleted lists.

- Don't digress, don't footnote, and don't wander. Keep each help file focused on a single, discrete task.

QUANTUM LEAP

Sometimes you must divide a lengthy help file into two separate help files. However, if you start to notice that many help files are too lengthy, online help may not be the proper medium for this information. Consider writing a traditional manual instead.

Beta-Test and Refine

Have real users test help files. Determine what topics they are searching for. Does your help system provide those topics? If so, do these topics appear in the table of contents and the index? Good online help systems are dynamic, adding, changing, or removing screens to meet customer needs.

Online Help Examples

Suppose a user wants to change the font of her favorite program. She clicks the Help button, which causes the program's online help system to spring to life. She then clicks the Index tab on the online help system and searches for the phrase *change font*. Almost instantly, the online help system displays the less-than-perfect help file shown in Example 11-2.

EXAMPLE 11-2   Online help (suboptimal)

Display Properties Screen

The Display Properties screen lets you control the display properties of your screen. In particular, you can use the Display Properties screen to control the following two properties:

- the background image displayed on the screen

- the font

You invoke the Display Properties screen by selecting File->Properties. The Display Properties screen appears, which looks like this:

Display Properties

Background Image	Font
⊙ Pastoral	⊙ Arial
○ Agrarian	○ Helvetica
○ Canyon	○ Times
○ Waterfront	○ Courier

[Apply Property Changes]

The program uses the background image in two places:

- in the splash screen that appears when you first start the program

- in the Build screen displayed when you request a build

To change the background image, select from one of the four choices and then click Apply Property Changes.

The program uses the font settings to display all its text. The Arial and Helvetica fonts are sans serif, the Times font is a serif font, and the Courier font is a fixed-width font. Select the desired font and then click Apply Property Changes.

The preceding online help file contains multiple problems. The poor reader simply wanted to find out how to change fonts. However, the writer seemed to be more concerned with describing all the options available on this screen. Consequently, the reader had to wade through quite a bit of extraneous information about other parts of the screen before hitting the part about changing fonts. To correct this problem, the writer should have created two help files—one on changing backgrounds and another on changing fonts.

The reader wants information quickly, but the preceding online help file makes it hard to find the relevant information. For example, the reader needs to know which menus to access. This information resides in the online help file, but it is diffuse rather than focused.

Finally, the information on fonts ("The Arial and Helvetica fonts are sans-serif") is fairly worthless in an online help file. Readers want quick recommendations.

The online help file shown in Example 11-3 provides improvements.

EXAMPLE 11-3   Online help (improved)

Changing Fonts

To change fonts, do the following:

1. Select `File->Properties`. The Display Properties screen looks as follows:

2. Select the desired font (from the highlighted section).

3. Select the `Apply Property Changes` button. Note that the program will make no font changes until you select this button.

We recommend the Arial or Helvetica fonts as they are the most readable online. The Times font is useful for printing lengthy screens since it looks nicer than Arial or Helvetica on paper. Courier is only valuable when accessing the built-in spreadsheet; otherwise, Courier is hard to read online and on paper.

Release Notes

Release notes are a compendium of information that didn't make it into a manual. To paraphrase Mark Twain:

> Release notes are a doc that everyone wants to have read but no one wants to read.

To the perpetual annoyance of customer-support organizations, very few customers actually read release notes. Those that do bother are rewarded with an intensely boring but very practical set of information. If you are writing release notes, assume that customers will search for nuggets but not read the whole thing.

Release notes typically consist of the following sections:

- an overview of the new features in this release

- a list of the known bugs, limitations, and issues in this release, along with any workarounds

- a list of all the bugs fixed in this release

- optionally, some additional documentation generated after manuals went to press

Release notes say a lot about your organization, which is why many organizations fight bitterly over what goes in them. The biggest cause of tension is typically the list of all known bugs, limitations, and issues. The tension is between those who say that listing a lot of known bugs speaks negatively about the quality of the product and those who say that customers appreciate knowing what's wrong. Resolve conflicts in face-to-face meetings rather than through e-mail wars.

Smart companies provide initial release notes with the product, then keep the release notes "live" on a Web site, making periodic adjustments throughout the life of the release.

The following page shows some sample release notes for a software product aimed at a technically sophisticated audience.

Be Careful about What You Put into Release Notes

Prospective customers sometimes look at the release notes to determine the number of known bugs it has and the company's commitment to fixing them. For marketing reasons, the release notes should show more bugs fixed than known bugs.

Release Notes Example: Carambola Web Server Version 3.7

These release notes describe version 3.7 of the Carambola Web Server.

New Features

This release introduces enhanced graphics caching. Our testing reveals a 20% to 35% performance improvement when serving pages that contain at least five images.

Bugs Fixed

Version 3.7 fixes the following bugs:

- **Bug #3581.** Users could not access the Cookie Dispenser during system backups. With Version 3.7, users who are members of the Systems group may access the Cookie Dispenser at any time.

- **Bug #3619.** Only one user at a time could access the Admin tool. With Version 3.7, an unlimited number of users (with the proper privileges) can access the Admin tool.

- **Bug #3924.** The Carambola Web Server did not work with the Dexco Proxy Server. With Version 3.7, the Carambola Web Server now works with the Dexco Proxy Server Version 2.5 or later.

Known Bugs, Issues, and Limitations

Version 3.7 contains the following known problem:

- **Bug #4273.** Enhanced graphics caching will not boost performance for users running Windows XP SP1.
  Workaround: Upgrading to Windows XP SP2 eliminates the problem.

Prefaces

Most writers follow one of two schools of thought regarding prefaces:

- Good manuals contain prefaces.

- No one reads prefaces, so why bother writing them?

I have a lot of sympathy for the latter opinion; however, a small percentage of people do read prefaces with an almost religious fervor. Therefore, it is a good idea to include a preface in every manual. However, since many readers skip the preface, it is also a good idea to repeat a few relevant facts from the preface in the opening pages of Chapter 1.

By custom, most prefaces answer the following questions for readers:

- **What is this book about?** Why might readers be interested in this topic?

- **Who is the target audience of this book?** What prerequisite knowledge or experience do readers require to understand this book?

- **How is the book organized at a chapter level?** Many prefaces provide a section that specifies the name of each chapter and a one- or two-sentence summary of its contents. (Always keep this section short since it is redundant with the table of contents.)

- **What is the pedagogic approach?** Explain why this is the best approach for explaining this technology.

Optionally, some prefaces also contain the following:

- a list of all "the little people" who made this book possible

- an explanation of why you wrote the manual

Having a Little Fun in the Preface

Some editors permit writers more freedom in the preface than in the rest of the manual. Consequently, writers who must conform to corporate restraints in the rest of the manual often bust loose in the preface like drunken sailors on shore leave. For writers who know what they are doing, an engaging preface can be a good way to hook people into reading the rest of the manual. Be careful though—if you lose readers in the preface, they won't read the rest of the manual.

Preface Example

The Carambola 4000 scanning electron microscope is the latest mid-level product from Carambola Industries. This microscope offers greater power and a simpler interface than previous electron microscopes.

This manual is aimed at readers who have at least a bachelor's degree in biology. Furthermore, it assumes that you are already familiar with basic microscopy techniques and terminology. Our research suggests that this model is most appropriate for scientists who already have at least one year of experience using an electron microscope.

After reading this manual, you will know how to use, troubleshoot, and maintain the Carambola 4000.

The manual consists of four chapters, which are organized as follows:

- Chapter 1 provides a tutorial to get you started with easy tasks.
- Chapter 2 explores the more advanced features of your microscope.
- Chapter 3 explains what to do if your Carambola 4000 stops working perfectly.
- Chapter 4 details routine maintenance procedures and cleaning techniques.

At Carambola Industries, we think that readers prefer to learn by using the product. Therefore, Chapters 1 and 2 show you how to use the Carambola 4000 to scan the sample slides that accompany the microscope.

I want to thank Mark for his knowledge of cell structure, Lauren for her amazing knowledge of electron microscopy, and Sal for his extraordinary mushroom pizza.

QUANTUM LEAP

As mentioned earlier, many readers skip over prefaces. For this reason, Chapter 1 should also define your audience. For example, you might consider providing a slightly modified version of the second paragraph (*This manual is aimed at...*) in Chapter 1 as well as in the preface.

Glossaries

A **glossary** is a small dictionary that defines the terms used in a particular document, Web site, or enterprise. When you are producing any form of lengthy technical communication—particularly those to which several people contribute—I highly recommend starting the project by creating a glossary. Getting contributors to agree on terminology upfront will save all sorts of time and aggravation later on.

The keys to writing a professional glossary are as follows:

- Keep definitions short.

- Provide relevant example usages in your definitions, where appropriate.

- Limit the terms in the glossary to primary concepts. Don't get bogged down defining too many secondary terms.

- Keep the form of each *term* grammatically parallel. Ensure that all terms are singular or that all terms are plural. However, if you pick singular, note that some terms are more natural in their plural form; do not force a natural plural into the singular just for the sake of consistency.

- Keep the form of each *definition* grammatically parallel. For example, do not define some terms with full sentences and others with partial sentences.

- Write unambiguous definitions. Make sure that the same definition could not apply to multiple terms. Compare and contrast terms.

- Target your audience appropriately. Define all the terms in the document that might be new or unfamiliar to your target audience. Don't define words that the entire audience already knows.

- Present terms in alphabetical order. Resist the temptation to organize terms by some other categorizing scheme.

- Provide hyperlinks to other terms within the glossary if this is an online document. In some cases, you may also provide hyperlinks to terms defined in other glossaries; however, make sure that the hyperlink opens the external glossary in a separate window. When referring to other glossary terms, use phrases such as *See also*, *Compare to*, and *Contrast with*.

- Aim to create a "closed-circuit" glossary in which any unfamiliar terms mentioned in definitions are, themselves, defined in the glossary.

The sample glossary on the next page is aimed at a well-educated lay audience. See also the glossary of technical writing terms that appears at the end of this book.

Glossary Example: Tropical Weather Terms

extratropical cyclone   An organized <u>low-pressure area</u>, formerly a <u>tropical cyclone</u>, whose energy source is the sharp temperature difference between warm tropical air and cold arctic air. Contrast with <u>tropical cyclone</u>.

eye   The center of a <u>hurricane</u>, which is usually marked by clear skies, low winds, and the lowest central pressure.

gusts[1]   The strongest transient wind speed. Gusts in a <u>hurricane</u> are typically 15% higher than the <u>sustained winds</u>.

hurricane   A <u>tropical cyclone</u> with <u>sustained winds</u> of at least 65 knots and an <u>eye</u>.

knot   A unit to measure <u>sustained winds</u> or <u>gusts</u>. A knot is equivalent to 1.15 miles per hour.

low-pressure area   A region marked by counterclockwise wind circulation (in the Northern Hemisphere), lower than average barometric pressure, and generally stormy weather.

sustained winds   The average wind speed as measured over a certain period (usually one minute or five minutes). Contrast with <u>gusts</u>.

tropical cyclone   A <u>low-pressure area</u> whose energy source is warm ocean water and whose <u>sustained winds</u> are greater than 25 <u>knots</u>. Contrast with <u>extratropical cyclone</u>.

tropical storm   A <u>tropical cyclone</u> having <u>sustained winds</u> between 35 and 64 knots.

---

1. Most terms are in singular form; however, *gusts* and *sustained winds* are in plural form because plural is more natural and more frequently used for these terms.

Tables of Contents

Technologists are busy people who rarely have time to read manuals linearly from cover to cover. Therefore, they often seek topics by searching the table of contents. For this reason, the table of contents is very important. You might be thinking, My word processor automatically generates the table of contents. How can I change it? Well, you enhance the table of contents by picking chapter and header titles wisely.

The following are some guidelines for creating a good table of contents:

- Think of all the first-level headers within each chapter as forming a list. Then, follow the rules for creating lists described in Chapter 7. In particular, use grammatically consistent header names to ensure parallel lists.

- Ensure that your header names contain key terms to attract the eyes of a browsing reader. For example, if you feel that readers are interested in mitochondrial DNA, then make sure that at least one of your chapters or headers contains this term.

- Keep chapter and header names relatively concise.

- In a lengthy manual, organize related groups of chapters into sections.

For instance, Example 11-4 contains a flawed table of contents that does not follow some of the preceding rules.

EXAMPLE 11-4 A Partial Table of Contents (with flaws)

The table of contents shown in Example 11-4 contains the following problems:

- The headers within Chapter 5 form a nonparallel list. For example, *Creating Tables* begins with a verb, while *Cells* is just a noun. The writer must rename the headers so that they all begin with a verb or all begin with a noun.

- The headers are missing several keywords that readers would likely seek. For example, many people reading this book would seek tag names in the table of contents. One of the headers—*Creating Headers (<th>)*—does contain a tag name; most of the other headers require them.

- The last header in Chapter 5 is too long. This header contains information that belongs in a paragraph, not in the header itself. In fact, the complexity of the header strongly suggests that this section be divided into two sections.

- The two chapters could be grouped into a section named *Tables*.

Example 11-5 shows a somewhat improved version of the table of contents.

EXAMPLE 11-5   A Partial Table of Contents (with improvements)

Indexes

Creating a good index is boring and time-consuming. Creating a great index requires supreme patience and diligence. Despite the pain, the results are extremely valuable to readers. After all, the majority of technical readers go straight to the index. If they can't find the magic word or phrase in the index, then that topic is lost to them, even if it is well covered in the manual.

Many people mistakenly believe that indexing involves highlighting selected words, kind of like students do with a yellow marker. In reality, good indexing is far more sophisticated. Even many professional writers feel they don't have the requisite skill or patience and turn over the task to trained professional indexers.

The Golden Rule of Indexing

When creating an index, pretend you are a reader, and ask yourself the following question:

What would I look up in the index?

All other indexing rules are subservient to this golden rule of indexing.

The following guidelines help you implement the golden rule of indexing:

- Create precise index entries; avoid vague entries.
- Avoid misleading entries that could send readers off on a wild goose chase.
- Permute index entries so that a single entry phrased as x,y is also listed under y,x.
- Provide index entries for related concepts, not just the literal words on the page.
- Use grammatically consistent forms for all entries.
- Don't forget to index the information appearing in graphics and tables.

How Long Should an Index Be?

It is hard to pinpoint the correct length for an index. After all, a single page can sometimes be adequately indexed with only two or three index entries, while a single juicy paragraph might require a dozen entries. As a broad estimate, I'd say that good indexes generally consume 5% to 7% of the length of the body of the book.

Indexes: Providing Concise Entries

To explore indexing, consider the following passage from a vegetable gardening manual:

> ## Adjusting Acidity
>
> Most vegetable plants prefer a slightly acidic soil with a pH between 6.0 and 6.8. A few vegetable plants, such as the tomato, prefer a slightly more acidic soil. The king of the acid lovers is the blueberry bush, which is happiest with a pH around 4.5.
>
> In most eastern parts of the United States, soil tends to be too acidic. The easiest way to raise the pH of soil is to spread crushed limestone. Over many western parts of the United States, soil is too alkaline. Spreading sulfur is an inexpensive way to lower the pH.

Some of the obvious index entries in the preceding passage are as follows:

> acidity
> alkalinity
> blueberries
> eastern United States soil
> limestone
> pH
> sulfur
> tomatoes
> vegetables
> western United States soil

The preceding entries are okay, but they are a little too general. The following entries are more useful because they are more specific:

> acidity of garden soil
> alkalinity of garden soil
> blueberries, proper pH of soil
> eastern United States soil, adjusting pH
> limestone, to adjust pH
> pH of garden soil
> sulfur, to adjust pH
> tomatoes, proper pH of soil
> vegetables, proper pH of soil
> western United States soil, adjusting pH

Indexes: Permuting Terms

The wise indexer permutes the words in some multiword index entries, making each appropriate permutation a separate index entry. For example, consider the following entry:

> acidity of garden soil

The preceding entry should lead to the following additional index entries:

> garden soil, acidity of
> soil, acidity of

You could also permute the following entry:

> eastern United States soil, adjusting pH

to entries such as the following:

> pH, eastern United States soil
> soil, eastern United States
> United States, eastern soil

Providing Second-Level Entries

Permuting entries invariably creates the need for second-level index entries. For example, keeping both of the following terms at the first-level is not optimal:

> soil, acidity of garden
> soil, adjusting in eastern United States

It looks more professional to keep *soil* as a first-level entry and to create second-level entries as follows:

> soil
>   acidity of
>   in eastern United States

Indexes: Providing Entries for Concepts

The golden rule of indexing yields the following less-than-obvious entries from the gardening passage:

> sweetening
> souring

Neither of these terms appears in the text, yet they are worthwhile index entries because many gardeners know them. (*Acidic soil is too sour and requires "sweetening."*)

Looking at a higher conceptual level, sometimes you have to sit back and ask yourself, what is this section all about? In many cases, the answer is found in the header. In other cases, you have to think beyond the obvious. For example, the following entries would also be worthwhile:

> crops, improving through pH balancing
> modifying, pH
> performance of garden, effect of pH
> yield of garden, improving through pH balancing

Parallelism in Index Entries

Indexes are a form of list. Therefore, you must apply the laws of parallelism to them. Most index entries are either nouns or verb phrases. The following simple rules should keep your index entries parallel:

- Place nouns in their plural form. Thus, specify *potatoes* rather than *potato*.

- Place verbs in their participle form. Thus, specify *creating* rather than *create* or *creates*.

Summary of Manuals

Before writing a manual, you should develop a complete doc spec and, possibly, a documentation project plan. Refer to Chapter 18 for details on planning, reviewing, and editing manuals.

While writing a manual, refer to Section 1 of this book to optimize all aspects of your manual. In addition, ask yourself the following questions:

- Does your preface (or Chapter 1, if you don't have a preface) explicitly identify who should be reading this manual and what the reader will get out of it?

- Do your table of contents and index contain listings for every keyword or concept that a reader might seek?

- Does the opening section of each chapter introduce the chapter's topic? Does the closing section of each chapter summarize the chapter's topic?

- Does your manual contain a nice mixture of text, tables, and graphics, or does one of those weigh too heavily? Generally speaking, manuals should contain a nice balance of words and pictures, although certain topics necessitate an unbalanced presentation. A manual of schematic diagrams probably won't contain too much text, and a manual on teaching scientists and engineers about technical writing probably won't contain a lot of graphics.

- In a glossary, are all terms and definitions grammatically consistent? Could the same definition apply to more than one term?

- Do the chapters follow naturally from each other, or do you need to reorganize them into a more logical order?

- Are any chapters too long? Divide lengthy chapters into two or three chapters. By the way, some writers feel that all chapters should be about the same length, but I've never heard a compelling argument as to why that would benefit the reader.

- Do certain chapters only have significance for a tiny percentage of the target audience? If so, they should become appendices.

While writing online help, ask yourself the following questions:

- Does your online help cover all (or nearly all) of the topics that readers seek?

- Is each online help section fairly discrete, or will readers be forced to read many different online help sections to answer their question?

- Can readers easily find the topics they seek?

 CHAPTER 12

Web Sites

The Internet is big. Really big. Important, too. Search engines have helped something or other immeasurably. The Internet has caused a fundamental change in yada, yada. In the time it took you to read this paragraph, 24 new Web sites were blah, blah, blah...

Having satisfied the legal requirements for all technical books published since 1994, I can now move on to the subject of this chapter, which is how to present technical and scientific information over the Web. This is not a chapter about HTML, JavaScript, or servlets. Rather, this is a chapter about designing and implementing Web sites that effectively convey technical or scientific information. Nevertheless, a little familiarity with basic HTML will come in handy as you read this chapter.

Is creating content for the Web different from creating content in a traditional paper manual? Well, clearly the Web and paper are different media. After all, the Web provides hyperlinks, which make it easy for users to bounce around like dust specks in Brownian motion. Furthermore, the Web offers multimedia, so Web sites can explain concepts through speech or music or animation.

In some ways, paper documentation has recently begun to act more and more like Web sites. For example, consider this very book you are now reading. Note how the book is chunked into bite-sized morsels, roughly the length of proper Web pages. Further note that you don't have to read this book linearly (from page 1 straight through to page n) to get something out of it. You can bounce around randomly as you would when browsing a Web site.

Television and the Internet have made us a society of information nibblers. There is no reason to editorialize against it, and the wise Web developer doesn't try to fight it. The wise Web developer simply embraces the notion that readers have Attention Deficit Hyperactivity Disorder.

Plans

A good plan improves your chances of Web success. This chunk provides some planning suggestions.

Define the Site's Purpose

Your plan should begin by identifying the purpose of the Web site, as in the following statement:

> **Web Site Purpose**
>
> This Web site will distribute recent scientific research on citrus trees and fruit to professional citrus farmers in Florida.

Define Your Audience

As with any form of communication, begin by defining your audience. All of the audience definition topics explored in Chapter 2 are also applicable to Web sites. Therefore, I recommend that you begin planning your Web site by filling out an audience definition chart similar to the one shown in Table 2-1. Supplement that table by supplying answers to the following additional topics for Web sites:

- How will users find your Web site? Typically, visitors will find your Web site through a search engine and from hyperlinks in a related Web site. If you expect visitors to find your Web site through a search engine, what keywords should trigger a search engine hit on your Web site?

- What Web sites will visitors hit *after* visiting your Web site? Perhaps your Web site will direct visitors to other Web sites, or perhaps a search engine will direct them.

View your Web site as one piece in your target audience's information jigsaw puzzle. Understand what piece your Web site will provide and what pieces other Web sites will provide.

Create a List of Pages

Your plan should contain a list of Web pages constituting your Web site. Many designers first create a list of topics and then herd these topics into Web pages. Other designers create a list of Web pages first and then figure out what information should be in each of them.

If your Web site will contain many Web pages, consider organizing Web pages hierarchically. The home page, of course, is at the top of the hierarchy. Just below the home page are the key secondary pages that you expect users to visit first. The secondary pages might lead users to tertiary pages, and so on. Table 12-1 helps organize the pages of a simple agricultural Web site.

TABLE 12-1   A Portion of a Web Page Planning Table for a Simple Agricultural Web Site

Parent Page	Web Page	Contents
Not applicable	Home page	Purpose and audience of site, navigator, search box, engaging picture, and About Us
Home page	Fertilizers	Summary table of various fertilizer categories, leading to tertiary pages
	Grafts	Summary table of different grafting techniques, leading to tertiary pages
	Infections	Summary table of various infections, leading to tertiary pages
	Pesticides	Graphic of various pests, leading to tertiary pages
Fertilizers	Organic	Table of organic fertilizers and relative efficacy
	Inorganic	Table of inorganic fertilizers and relative efficacy
Infections	Citrus Canker	Summary of latest research; links to key sites
	Nonbacterial	Table of nonbacterial infections

Pick the Right Web Technologies

Your plan should consider technological questions such as the following:

- **What parts of your Web site will provide static information and what parts will be dynamically generated?** *Static content* looks the same to every visitor on every visit. By contrast, *dynamic content* can vary per visit, based on factors such as who the visitor is (this is called *personalization*) or what the visitor wants to do.

- **Who may access your Web site?** Most technical Web sites provide free access to everyone, although some do restrict access for security or financial reasons.

- **What technology will you use to evolve the Web site?** If a Web site consists of only a few pages, then no supporting technology is necessary. However, as a Web site grows past even a small number of pages, maintaining the Web site becomes difficult. To untangle the mess, consider content-management software.

Home Page: Specify Purpose and Audience

The **home page** is the designated starting page for a Web site. It is the most important page at any Web site and the one you should devote the most time to implementing. This chunk and the two that follow provide guidance on building home pages.

The first order of business with any home page is to state clearly the purpose of the site and its audience. Unlike a commercial site, scientific and technical sites do not need a catchy slogan or a cute catchphrase. Instead, such sites should just blurt out the purpose and audience in the following three places:

- **In the title of your home page.** Use the HTML <TITLE> tag to specify the title. The title appears in your browser's title banner, in browser bookmarks, and in search engine results. Keep the title brief—no longer than eight words.

- **In the top-level header on your home page.** You typically specify the top-level header through an <H1> tag.

- **In the opening body text of your home page.** The opening body text is typically inside a <P> tag, but it might just as easily be inside a <TD> tag.

For example, the following HTML passage shows the start of an agricultural Web site:

```
<HTML>
<HEAD>
   <TITLE>Latest Citrus-Growing Research for Farmers</TITLE>
</HEAD>
<BODY>
<H1>Latest Citrus-Growing Research for Farmers</H1>
<P>This site details the latest agricultural research on growing oranges,
lemons, and limes. We've built this site for professional citrus farmers,
particularly those with agricultural degrees.</P>
...
```

The Title

Writing a good title is a difficult balancing act because a good title is simultaneously concise and precise. Somehow, you must capture the essence of your Web site in a bare minimum of words. As a rule of thumb, do not use more than eight words in a title. However, before congratulating yourself on the slimmest possible title, remember that a title that is too general is pretty worthless as well. For example, the following title is too general; based on the limited information it presents, the Web site could just as easily be for diabetics, doctors, or researchers:

```
<TITLE>Artificial Insulin Research</TITLE>
```

By contrast, the following title precisely identifies the audience:

```
<TITLE>Artificial Insulin Research for Type B Diabetic Patients</TITLE>
```

Title and Search Engines

When displaying matching Web sites, search engines display the exact title embedded in your <TITLE> tag. Consequently, a precise title attracts appropriate visitors to your site.

Home Pages: Engage the Reader's Imagination

Surfing the Web is similar to cruising for a television show. In either case, potential viewers decide whether new content is worthwhile in a handful of seconds since alternative content is only a click away. (Rest assured that this great big World Wide Web almost certainly harbors other Web sites that cover similar territory to yours.) The moral: a good home page makes its case very quickly.

Presenting an eye-catching graphic is probably the best way to engage readers in a home page. The key is to pick a graphic that is appropriate for the audience. (See the next page for details.)

Another more devious mechanism—one only appropriate for lay sites—is to pose a few mysteries that can only be answered by exploring the Web site. This mechanism is effective because intelligent readers have trouble passing up a challenge. For example, the portion of a home page shown in Figure 12-1 features questions about Saturn. Notice that each question on the home page is a hyperlink to the answer.

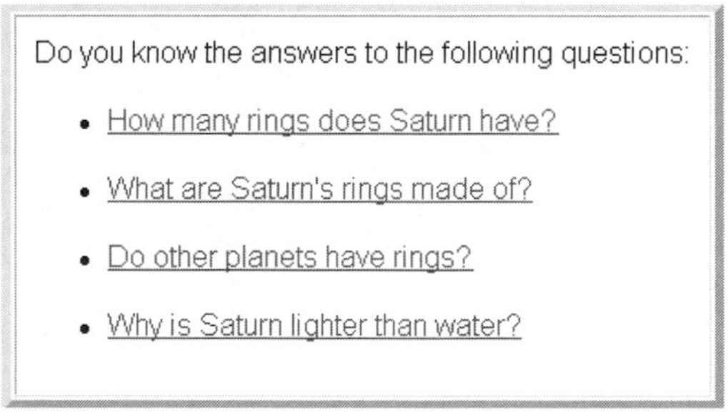

Do you know the answers to the following questions:

- How many rings does Saturn have?

- What are Saturn's rings made of?

- Do other planets have rings?

- Why is Saturn lighter than water?

FIGURE 12-1   Engage lay readers by posing questions on the home page.

You may scoff at this technique, thinking that it sounds a bit childish. However, this technique is effective because it gets users to leave the home page and start probing the site's real content. Remember that this technique is inappropriate for a serious academic Web site.

Home Pages: Set the Tone

A well-constructed home page should provide obvious visual clues enabling visitors to identify the Web site's tone immediately and to determine whether the site is appropriate for them. In essence, setting the tone is similar to defining the audience explicitly through text, although setting the tone acts at a more emotional level.

For example, consider home pages devoted to the planet Saturn. For a Web site aimed at a lay audience, the home page should feature a large, beautiful, color photo of the planet. By contrast, for a Web site aimed at planetary scientists, the home page might feature a graph or perhaps a map of magnetic fields on the surface.

A Site for Technical Practitioners

A serious Web site aimed at serious researchers requires a serious tone. The following home page guidelines help set that serious tone:

- Use common sans-serif fonts such as Arial or Verdana, which are usually the browser's default fonts anyway. Never use exotic fonts. Render text in black on a white background.

- Provide an overtly technical illustration on the home page.

- Make sure that text on the home page contains at least some jargon.

- Provide a fairly high amount of information on the home page, although not to the point of clutter.

A Site for a Lay Audience

A technical or scientific Web site aimed at a lay audience requires a gentler, less-intimidating foyer than a site for technical practitioners. For example, the home page for a site that explains asbestos to a lay audience should not display an X-ray of a diseased lung. Many nonscientists find science intimidating or humbling. Therefore, your home page needs to provide a comforting environment, much like the waiting room of a doctor's office. The following list provides a few guidelines for setting the tone:

- Provide some color.

- Provide a photograph that contains a person interacting with the target technology. (Ideally, the person should be smiling. If smiling is inappropriate for the subject matter, at least show the person in apparent mastery over the technology.)

- Do not use jargon.

- Provide plenty of white space in the design, giving the home page a relaxed, airy feel.

Page Templates

Most of the Web pages within a single large commercial Web site share the same layout. For example, although the Web site for the online edition of a large newspaper might contain thousands of pages, 90% of them might share the layout shown in Figure 12-2.

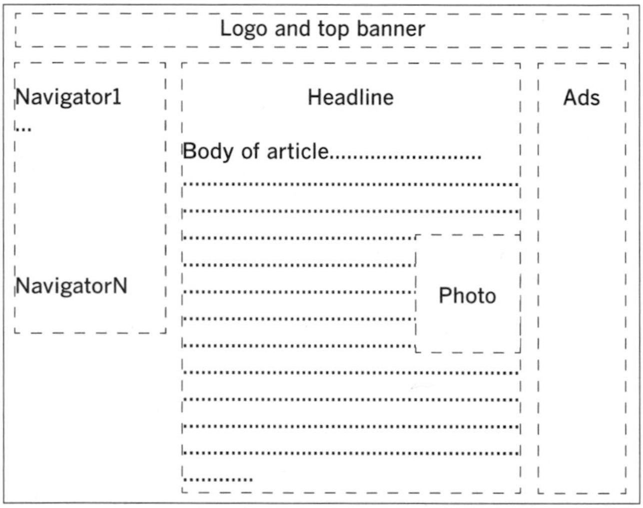

FIGURE 12-2   Layout of a typical online newspaper.

To implement a consistent layout such as the one shown in Figure 12-2, you must define a **page template.** A page template defines the layout elements of a class of Web pages. When a Web site uses page templates consistently, readers quickly learn where the content is, where the navigational aids are, and even where to avoid looking (at the ads, I suppose). A site without page templates forces visitors to regain their bearings whenever they step into a new Web page. A Web site composed without page templates is like a program whose graphical user interface changes with every use.

Page templates are not just for large, commercial Web sites. Even a modest, technical Web site with only a dozen Web pages should still rely on a page template. A simple page template, when applied to all pages in a Web site, will give a site a far more professional look.

You can use a variety of techniques to implement page templates. The most sophisticated page template mechanisms are built into enterprise content-management software. Such software enforces the use of page templates, making it impossible for renegade cowboys to create different-looking Web pages.

Figure 12-3 shows two different Web pages created from the same page template.

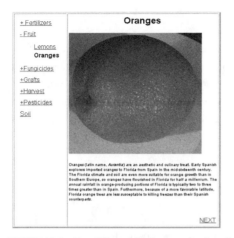 

FIGURE 12-3   Two different Web pages rendered with the same page template.

A Set of Page Templates

A large Web site typically contains multiple page templates. Each page template represents a different variety of pages. For example, you might have one page template for lab reports and another page template for somewhat more casually presented information. However, even in Web sites that use multiple page templates, all the page templates usually share some common themes. For example, all page templates might share the logo of the organization that supports the Web site.

Navigators and Search Boxes

Each page in a Web site should provide a pathway to reach all other pages in your Web site. That is not to say that each page must provide a one-click link to every other page. (That much navigational baggage would crush any large Web site.) Nevertheless, each page should provide a clear pathway allowing readers to move to any other page in the Web site in just a few clicks. The most popular strategies for navigation within a Web site are as follows:

- navigators

- search boxes

Whichever method you choose, be consistent about their placement. For example, if a search box appears in the top-right corner of your home page, then that same search box should appear in the top-right corner of every page in your Web site. Don't make readers hunt for navigational devices.

Navigators

A **navigator** is a list of multiple hyperlinks (and nothing but hyperlinks) that helps readers get to the desired page. For example, consider the simple navigator in the agricultural Web site shown in Figure 12-4.

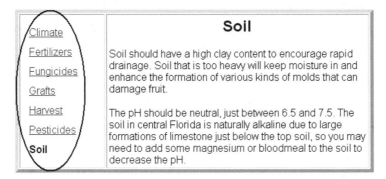

FIGURE 12-4   The left side of this Web page contains a simple navigator.

The navigator should clearly distinguish the current page. For example, since Soil is the current page, notice that the navigator does not display a hyperlink for Soil.

Notice that the anchor text (for example, Climate, Fertilizers, and so on) for each hyperlink in Figure 12-4 is only one word long. Your own navigator can contain more than one word, but one word is certainly the ideal.

The navigator for a simple Web site can contain a pointer to every page in the Web site. However, once a Web site contains more than 15 or so pages, a navigator with links to every page starts to become ungainly. (By the way, try to avoid forcing readers to scroll down to see a submerged part of the navigator.) To keep a navigator at a manageable size, consider a multilevel navigator. For example, if the reader clicks Climate, appropriate additional choices should appear, as shown in Figure 12-5.

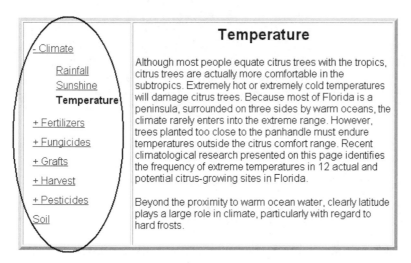

FIGURE 12-5   A two-level navigator, which is typically implemented through JavaScript.

Search Boxes

Instead of searching for the right hyperlink in a menu or navigator, users love to type text into a search box and let it find the right link. To be more precise, most readers start off loving search boxes but end up frustrated with them because of the spurious links they generate. Nevertheless, if you do implement a search box, make sure that you clearly define which site(s) will be searched. The simple phrase *Search* is no longer clear; it could mean *Search this Site Only*, *Search the Entire Internet*, or anything in between. By contrast, Figure 12-6 shows a well-defined search box.

FIGURE 12-6   An unambiguous search box.

Hyperlinks in Body Text

During the Web's infancy, many users thought that hyperlinks were its coolest feature. Nowadays, Web users take this critical technology for granted. This chunk explores how to get the most out of hyperlinks within body text.

Hyperlinks Are Revolutionary

Hyperlinks represent a sea change in how people search for technical information. True, hard-copy books have long contained endnote citations and bibliographies, which, like hyperlinks, point readers to similar sources. However, unless readers happened to be in a major library, access to cited works could be weeks away. By contrast, modern scholars now complain when a hyperlink forces them to wait more than ten seconds.

Click Here

Web designers hate using the phrase *click here* as anchor text. For example, approximately 100% of Web designers believe that the following hyperlink looks amateurish:

> To see Dr. Jill Black's tutorial about eczema, click here.

Web designers greatly prefer using the target word or phrase as the anchor text. For example, the following example omits *click here* altogether:

> Dr. Jill Black created a tutorial about eczema.

Once in a while, it is clearer to write a passage with *click here* than to adhere dogmatically to the *click here* prohibition. In fact, if your Web site's target audience includes many relative newcomers to the Web (yes, this still happens), then *click here* is often the least ambiguous choice.

QUANTUM LEAP

In this age of computer viruses, many readers are justifiably nervous about clicking hyperlinks unless they are sure of what they are getting into. When a user clicks a hyperlink in your Web site and is transported to a surprising location (even a benign one), that user often becomes suspicious about the rest of your Web site. All anchor text should honestly and accurately represent the target URL.

The Right Number of Hyperlinks in Text

Too many hyperlinks can confuse readers. For example, consider the following hyperlink-happy sentence:

> Dr. Jill Black, a renowned dermatologist, created a tutorial about eczema, one of many common skin diseases that can be triggered through allergens or heredity.

The preceding sentence contains six hyperlinks, which is an overwhelming density. If readers followed every one of these, it could take them hours just to process that one sentence. In addition, since hyperlinks are usually rendered in the default blue color, the preceding black-and-blue mixture takes on a striped appearance that is somewhat difficult to read.

A reasonable rule of thumb is to keep hyperlinks down to one per sentence, although even one per sentence applied across any entire page of text would be way too many.

Default Blue Hyperlinks Are Best

By default, browsers render hyperlinks in blue and underline the anchor text. It is fairly easy to change this default; however, I recommend against it. Changing the default forces readers to slow down to figure out what graphic notation represents a hyperlink.

A Separate Target Window

When visitors follow hyperlinks, they typically jump off the current Web page. True, they might arrive at a different part of the same Web site, but they still lose context and drift further and further away from the original Web page, eventually waving it good-bye. For this reason, the clever Web developer considers implementing hyperlinks so that the target appears in a separate window. That way, visitors can stay on the current Web page while exploring other hyperlinks.

Be careful how you make additional browser windows appear, lest you trigger a pop-up blocker's wrath.

Secondary Pages

As noted earlier, all secondary pages (that is, nonhome pages) should use a page template to present a consistent format. This chunk provides a few suggestions about those secondary pages.

Keep Content of Each Page Relatively Brief

Keep the amount of content on each Web page relatively brief. In the 1990s, a design axiom stated that users hate to scroll, so Web pages should be short enough to fit in a standard screen without a scrollbar. However, in the last few years, users have gradually become accustomed to vertical scrolling, and the design prohibition against vertical scrollbars has ended. Nevertheless, lengthy Web pages (beyond about four scrolled pages) are still frowned upon.

For example, suppose you must convert this book into a Web site. As a good Web developer, you would make a Web page out of each one- or two-page chunk rather than putting an entire chapter into a single Web page.

Associate Related Web Pages

Since content must be brief on each Web page, you need to find a way to associate related Web pages. Probably the best way to do this is through a navigator (described earlier in this chapter). In addition, you might provide See Also hyperlinks at the bottom of each Web page. Since the Web is, by its nature, a random-access medium, you would generally not put Next buttons at the bottom of pages unless each page is truly meant to be read in a specific order. For example, it would be appropriate to put Next buttons at the bottom of tutorial pages.

Keep Content Narrow

Although users have become accustomed to scrolling vertically, most users despise scrolling horizontally. (Note that most modern mice provide a vertical scroll wheel but not a horizontal scroll wheel.) Furthermore, placing text in relatively narrow columns helps users read quickly. Therefore, the wise Web developer follows the same practice as most online newspapers and keeps body text between 350 and 500 pixels in width.

Graphics—particularly high-resolution photos—are often rather wide. Within a Web page, avoid displaying wide graphics that will force users to scroll horizontally. Instead, display a scaled-down version of the graphic and provide readers with a hyperlink that will pop the full version up in a separate window. (Beware of pop-up blocks, though.)

Text in Web Sites

Most of the rules established in Section 2 of this book also apply to Web pages. In other words, your Web content should consist of short, clear, active-voice sentences that use vocabulary appropriate for the target audience. However, creating content for the Web is not exactly like writing traditional documentation. When creating content for the Web, the wise writer considers the following differences:

- The "Tell 'em what you're going to tell 'em... tell 'em... tell 'em what you told 'em" approach is less valuable in Web writing than in traditional writing. Web visitors frequently don't have the patience to read introductions and seldom have the patience to read conclusions. Web visitors typically just want to dive straight into the entree, skipping the appetizers and dessert.

- You have a range of media at your disposal. Some kinds of information are more effectively presented through audio or video than through text. In fact, Web visitors tend to have less patience for straight text than do readers of hard-copy documentation.

- To handle a long block of text within a Web page, consider the following approach:

    - Use subheads to divide the long block of text into units.

    - Start the Web page with a bulleted list.

    - Make each element in the bulleted list a hyperlink, pointing to subheads within the page. The Web page skeleton shown in Figure 12-7 illustrates this approach.

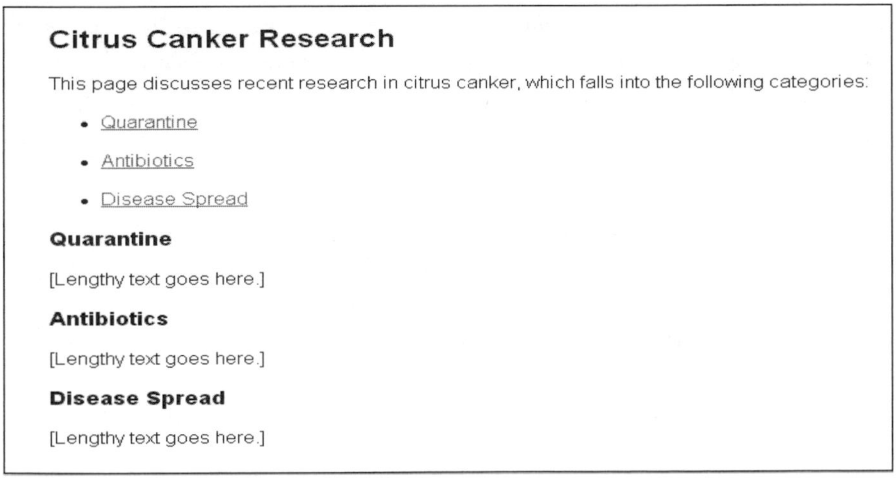

Citrus Canker Research

This page discusses recent research in citrus canker, which falls into the following categories:

- Quarantine
- Antibiotics
- Disease Spread

Quarantine

[Lengthy text goes here.]

Antibiotics

[Lengthy text goes here.]

Disease Spread

[Lengthy text goes here.]

FIGURE 12-7   Handling large blocks of text through bulleted links and subheads.

PDF versus HTML

On the Internet, the two most common document distribution formats are HTML and PDF. Table 12-2 compares and contrasts these two formats.

TABLE 12-2 Comparing PDF and HTML documents

Question	PDF	HTML
Can the user change an author's font choices?	No. The user must live with the author's font choices.	Yes and no. The answer depends on how the author has set up the document.
Can the author embed fonts?	Yes. The author can optionally embed fonts within a PDF file to ensure that users can view the document properly.	No. The author cannot embed fonts within an HTML file. The author can, however, suggest alternate fonts to use if the author's first choice of fonts is not available on the user's machine.
Does text rejustify when the user resizes a window?	No. When the user resizes the Acrobat Reader window, text does not rejustify.	Yes (usually). When the user resizes the browser window, text typically does rejustify.
Does font size change when the user resizes a window?	Yes. When the user resizes the Acrobat Reader window, virtual font size adjusts. Thus, fonts look relatively small when the viewing window is small.	No. When the user resizes the browser window, font size does not change.
Does this format use the concept of pages?	Yes. PDF uses the traditional physical page metaphor, just like a word processor.	No. A user can scroll almost infinitely down a lengthy document without encountering page breaks.
Do documents print well?	Yes. Printed pages look almost exactly as they do when printed directly from the author's word processor.	No. Pages print in strange and uncontrollable ways. Sometimes information on wide lines doesn't get printed at all.

PDF documents look awfully similar to a traditional book. PDF gives authors *complete control* over the layout of each page, just like WYSIWYG word processors. With HTML, the author hands over some of the formatting power to the browser and some to the end user. PDF is rigid; HTML is flexible.

Should You Distribute PDF or HTML?

In an ideal world, you would distribute documents in *both* PDF and HTML format. After all, some users prefer PDF and some prefer HTML. However, distributing in both formats is a lot of work.

Many users still prefer hard-copy documentation to soft-copy documentation. A few of these users have such a strong preference that they print out online documentation. Since the hard-copy version of PDF documents looks much better than the hard-copy version of HTML documents, score a point for PDF.

The average author typically converts a lengthy manual into a set of *multiple* HTML files; however, that same average author typically converts that same lengthy manual into a *single* PDF file. The PDF version often contains several orders of magnitude more bytes than one HTML file. Consequently, if your users access online documentation over a low-bandwidth network, HTML is a better choice.

Finally, some users read documents from devices (such as most cell phones) that cannot display PDF. However, a rather extraordinary collection of devices (not just traditional computers) can display some form of HTML.

Programs to Handle PDF

Adobe invented PDF; however, PDF is an open specification, and some other vendors have created software products for interacting with PDF. Nevertheless, the key software programs for interacting with PDF are still made by Adobe.

Acrobat Reader is a program that displays PDF files. It is freeware, available from www.adobe.com. Many modern computers preinstall Acrobat Reader. Furthermore, an Acrobat Reader plug-in lets users view PDF files directly in their browsers.

Acrobat Distiller is a program that converts documents to PDF. The full-strength version of Acrobat Distiller (generally marketed within a larger package called Acrobat Professional) lets authors convert just about any kind of document (including, ironically, HTML files) into PDF. Some word processors embed a reduced-strength version of Acrobat Distiller that enables conversion of only certain kinds of documents into PDF.

Adobe (and others) also make products that enable authors and others to edit PDF files. However, instead of editing the PDF file, it is generally far smarter to edit the source document, and then regenerate the PDF file.

Frequently Asked Questions (FAQs)

This chunk details frequently asked questions (FAQs), which are a staple of technical Web sites.

What is a FAQ?

A FAQ—as you probably know—consists of a bunch of questions and their answers. In addition to Web sites, newsgroup snobs love to use FAQs to intimidate newcomers. (*Why didn't you read the FAQ before posting that question?*)

Why do many readers hate FAQs?

Most readers like the concept of FAQs; however, the actual FAQs often leave something to be desired. Usually, readers are angry because FAQs don't contain *their* questions.

How should you create a FAQ?

The technique for creating a FAQ is simplicity itself—just do the following:

1. Gather up a list of the most commonly asked questions.

2. Provide answers for each of them.

3. Organize groups of related questions into sections.

4. Optionally, provide a search mechanism to help readers find the best-fit question.

How do you gather up a list of questions?

For product-related FAQs, just ask your customer-support organization, since they usually keep records of all the questions that customers ask. For science or general technology FAQs, questions posed on newsgroups and electronic bulletin boards often provide good sources of questions.

Can you make up your own questions?

Absolutely. Don't rely completely on questions that others have already asked. Sometimes people are too embarrassed to ask questions that they think are too basic. You, the writer, should make up some of the questions. The best FAQs successfully anticipate readers' questions.

What questions should you ask first?

The first question should always explain the purpose of the FAQ. The next few questions should summarize the product or technology. In essence, you should write an introduction in Q-and-A format.

Should you include every possible question in a FAQ?

No. A question asked only once does not constitute a frequently asked question. Don't clutter the FAQ with arcane questions.

What is the key to answering questions?

Be concise; answer the question in a paragraph, if possible. Never write an answer longer than three paragraphs. If an answer can't fit in three paragraphs, split the question into two or three narrower questions.

Don't try to explain the whole world in each answer; create hyperlinks and cross-references to other related answers that might already have the goods.

What is the proper tone for answers?

Readers do not expect a formal tone in FAQs and will appreciate a casual, down-to-earth approach. Writers can often let their hair down in a FAQ.

Can you place graphics in answers?

Yes—readers love graphics.

Is humor okay in a FAQ?

Yes, but remember that it is notoriously difficult to create a response that everyone in a diverse audience will find humorous.

If grouping questions, how many questions should you place in each group?

Keep groups of questions fairly small (fewer than 10 questions). When groups of questions get too long, subdivide them into narrower categories.

How do you make a FAQ searchable?

Most large Web sites provide automated indexing, so a FAQ placed in such a Web site will automatically be searchable. If your FAQ does not have the benefit of automated indexing, consider placing the entire FAQ in one physical file. Readers can then invoke their browser's Find menu to search for key words. The only problem with this approach is that a single file FAQ—particularly one with a lot of graphics—might take an annoyingly long time to download. To get around this problem, consider placing only the questions (not the answers) in one physical file. Then, place each answer in a separate file. Users can download the list of questions quickly, since the question file is usually relatively small and typically does not contain graphics.

Summary of Web Sites

The home page is the most important page in your technical Web site and the one that you should expend the most energy on. When reviewing your home page, ask yourself the following questions:

- Does the home page clearly explain the purpose of and audience for the site?

- Does the home page engage the reader's imagination and encourage exploration?

- Does the home page establish the tone for the site?

- Does the home page provide a mechanism for navigating and searching the rest of the site?

When reviewing your secondary pages, ask yourself the following questions:

- Is the writing accurate, concise, and audience appropriate? Does the writing conform to the best practices highlighted in Section 2 of this book?

- Do the secondary pages share a common layout, enforced through page templates?

- Does each secondary page contain a descriptive top-level header and title?

- Is the body text on any secondary page longer than four scrollable pages? No secondary page should be longer than four pages. If you find a Web page that is too long, divide it into multiple pages.

- Are any pages too wide? Pages shouldn't force users to scroll horizontally.

When reviewing your entire Web site, ask yourself the following questions:

- Does your Web site truly live up to the purpose and intended audience stated by the home page?

- Can visitors easily find their way around the different pages in your site?

- Can visitors find your site from a search engine?

CHAPTER 13

Proposals

A proposal is a written request to do something—perform research, start a company, develop a program... you name it—in exchange for money or other resources. Proposals are a large part of the academic, research, and business worlds. The following are some of the more common kinds of proposals that engineers and scientists write:

- **Research proposals.** These are requests to organizations such as the National Science Foundation or European Organization for Nuclear Research (CERN) to fund scientific research. Research proposals are the economic lifeblood of many universities and sciences labs.

- **Business plans.** These are requests to venture capitalists (VCs), banks, or other investment houses to fund or expand a company.

- **Book proposals.** These are requests to a publisher to fund a book.

At first glance, this may seem like a disparate group; however, all proposals have a lot in common. Beyond the obvious similarities in content (for example, cover letters, abstracts, and bios), the purpose of any proposal is to convince the target audience that you are exactly the right choice for an investment. Therefore, understanding what your target audience wants is critical. This chapter details all three types of proposals but first explains some general proposal-writing concepts.

Are Proposals Truly Important?

Engineers who believe that proposals are unimportant are as successful as actors who think that physical appearance is overrated. If you learn to write effective proposals, money and other resources will flow out of the tap and into your cup. If you stubbornly insist that good ideas are more important than good proposals, your cup will remain dry, and you will turn bitter. The goal for this chapter is to improve your chances of getting proposals accepted.

The Proposal before the Proposal

Before writing a formal proposal, your team should produce a **preproposal**, which is a brief, clear summary (oral or written) of the project. Producing this preposal is often agonizing since team members will likely have divergent ideas at this stage. Nevertheless, it is important that the team reach consensus fairly early. Each member of the team should become fluent in the preproposal and be prepared to pitch it on short notice.

Creating formal proposals is a time-consuming, agonizing affair for most writers. Before investing this kind of effort, you typically need to get some sort of initial affirmation. For example, prior to writing a book proposal, you should present the basic ideas for the book in a short conversation with a publisher. If the publisher likes the preproposal, he or she will sanction a formal book proposal. Some publishers won't even read a formal book proposal unless they've first read or heard a preproposal.

Elevator Speeches

A preproposal for a business plan typically takes the form of an **elevator speech**. Imagine that you find yourself standing next to a venture capitalist (VC) in an elevator. Can you pitch a compelling business plan before the VC gets off? You only have 60 seconds, so here's what to say:

- Introduce yourself and state your credentials. (*Hi, I'm Kay Linda, president of Dexco Unlimited.*)

- Describe the problem that your company plans to solve. To put it another way, describe the potential market. (*Did you know that 10% of the adult population suffers from some form of heartburn?*) Ideally, you should get the VC to identify with the problem.

- Describe your company's amazing solution to the problem. (*I've developed an over-the-counter antacid that reduces acid 12 times more effectively than calcium carbonate.*)

- Explain in nice round numbers and percentages how much money you need and what you plan to do with it. (*We need four million dollars, of which approximately 50% will go into completing research, 30% will go into marketing and distribution, 10% will go into fabrication, and 10% will cover salaries and expenses for 18 months.*)

VCs who leave the office for a while often return to a voice mailbox full of elevator speeches. The best speeches convince VCs that a market is waiting to be tapped. As with anything business related, remember to focus on the financial part of things and to minimize the technical wizardry.

Adherence to the Proposal Template

Most proposal committees provide standard proposal templates or guidelines. You need to follow the proposal guidelines as closely as possible. If the proposal requires that you complete 132 sections—half of them clearly for worthless bureaucratic reasons—then you must complete all 132 sections. Do not add extra sections. Do not delete sections. Do not ignore sections. Creativity is a wonderful thing, but this isn't the place for it.

What exactly does *following the template* mean? Well, suppose the proposal template asks you to provide the names of three customers currently using your product. As an over-achiever, you think to yourself, "Heck, if three customers are good, then four or five ought to be even better." In this case, though, three really is better than four or five. Similarly, if the template tells you to keep the "Technical Background" section under 250 words, then don't feel that providing 500 words is twice as good. If the "Work Plan" section requires a Gantt chart, then you must provide a Gantt chart even if you already have a much better schedule inside a spreadsheet.

The Consequences of Not Following the Template

Suppose you are a member of the proposal review committee. Your committee must distribute a ten-million-euro grant for a four-year project. Two proposals arrive, both containing brilliant ideas, but only one of them follows the template. Your committee picks the conformant one. Why? The winning proposal team proved a willingness to follow the committee's rules. The losing team did not. The committee had no intention of entering into a four-year agreement with a team that would be difficult to work with.

Proposal templates are not always clear. What if you don't understand some aspect of the template? You need only ask. For research proposals, the reviewing committee typically designates a program manager whose duties include talking with submitters. It is less embarrassing to ask questions up front than to submit a confused proposal. In addition, the proposal review committee often publishes copies of successful proposals. Emulate the winners.

QUANTUM LEAP

What proposal review committees see in a proposal is what they think they'll get in the proposer. Make sure that your proposal is tidy and typo free.

Proposal Element: Cover Letters

Most proposals require a **cover letter**, which introduces the proposal and explains why it was sent to the recipient. A cover letter should be short—generally only three paragraphs or so. Where appropriate, it should remind the reader of previous interactions. If appropriate (for example, in a business plan's cover letter), consider providing a "drop-dead date," identifying what will happen if the recipient does not respond by a certain date.

Example 13-1 shows a sample cover letter for a business plan.

EXAMPLE 13-1   Sample Cover Letter

```
                                        Attn. Barney Rutherford
                                        Dexco Unlimited
                                        317 Kenyon Ave.
                                        Palo Alto, CA 94306
                                        1-650-555-1212
                                        br@dexcounl.com
                                        May 1, 2005
Monadnock Investments
25 Hewitt St.
Cambridge, MA 02139
Attn: Janet Mertrie

Dear Ms. Mertrie,

When we met last month in Atlanta at EntCon, you asked me to prepare a formal
business plan for an initial round of funding at Dexco Unlimited. I'm pleased to
provide the attached business plan, which explains Dexco Unlimited and why we'd
like additional funding.

As you probably recall, Dexco Unlimited has created a prototype for a new
enterprise software application that provides personalized human resources
information. We believe that our application will dramatically reduce the total
cost of ownership for small and medium-sized enterprises and compete effectively
against existing firms in this space.

Dexco Unlimited is delighted to offer this initial funding opportunity exclusively
to Monadnock Investments. If Monadnock Investments does not respond by June 1,
2005, Dexco Unlimited will present this business plan to other potential
investors. If I can answer any questions, please do not hesitate to contact me
directly.

  Sincerely,

  Barney Rutherford
```

Proposal Element: Biographies

Proposals require a short biography (bio) of all the principals, possibly with résumés or curriculum vitaes (CVs) attached. When writing a biography, remember that the goal is to convince your target audience that you are exactly the right person to receive the grant. Unfortunately, many people find writing a bio one of the most daunting parts of the entire proposal writing process, and are scared off by the following two conflicting thoughts:

- My background isn't fancy enough to impress the proposal reviewers.

- It isn't right to brag.

Let me end the conflict for you—it is not only right to brag, it is essential. In fact, it is detrimental to think of this as *bragging*; instead, think of it as optimizing your team's chances of winning a proposal competition. Besides, it ain't braggin' if it's true, so make sure that everything in your bio is 100% true. Every fact on your bio must pan out if the proposal reviewers do a check. Lying on your bio is one of the quickest ways to eliminate yourself from contention.

> **Half-Lies**
>
> Don't write bios that are technically true but misleading. For example, suppose that you played an administrative role on a team that made a major breakthrough. Your bio should not vaguely describe yourself as being a "member of the team." Instead, your bio should describe your specific role, such as "administrative assistant on the team."

Take a look at the bios of people who have succeeded with the proposal review committee. What do these bios have in common? Is it a shining education? Is it technical leadership experience? Is it experience running small businesses? Did previous winners emphasize their publications?

Experienced people often have trouble deciding what to emphasize in a short bio. Don't use the same stock biography for each proposal; cater it for each audience. For example, consider an experienced technologist who is creating a business plan aimed at venture capitalists. The goal of venture capitalists is to make money. Therefore, the following biography would be suitable because it emphasizes management experience:

Biography of Rajendra Priyadarshan (appropriate for a business plan)

Rajendra Priyadarshan has managed software organizations for the last eight years. For the last four years, Rajendra has served as the director of engineering at Dexco Unlimited, where he manages a staff of 23 and an annual budget of over three million dollars. His engineering team produces two major and four minor product releases annually. Before coming to Dexco, Rajendra managed a QA team for four years at Carambola Software; during this period, the company earned a J. D. Power quality award. Rajendra also has worked at a variety of other organizations including MIT, Digital, and Prime Computer. He holds an MBA from MIT's Sloan School and a BS in computer science from Rensselaer Polytechnic Institute.

On the other hand, if Rajendra needed a biography for a book proposal on a particular technology, it would be wiser to emphasize literary and technical experience over managerial experience. In fact, too much managerial experience might suggest to reviewers that Rajendra is out of touch with technology. Consider the following biography, attached to a proposal to write a book on software engineering in the Java programming language. This biography aims to convince the target audience that he has the appropriate experience to write a successful book. Notice how this biography begins by emphasizing Rajendra's publication and technical experience.

Biography of Rajendra Priyadarshan (appropriate for a book proposal)

Rajendra Priyadarshan writes the popular "From the Trenches" column for Software Engineering Monthly. He has led teams of Java programmers on enterprise software applications since 1995. With more than 20 years high-tech experience at companies such as Dexco Unlimited and Carambola Software, Rajendra carries a wealth of writing, programming, and management experience. He holds an MBA from MIT's Sloan School and a BS in computer science from Renssalear Polytechnic Institute.

QUANTUM LEAP

Some business plans now include bios not only of principals but also of the company's key consultants. Attaching your company's name to a "star" often adds clout to a business plan.

Proposal Element: Abstracts

Nearly all proposals require an abstract or summary. An **abstract** is a highly concise summary of your proposal. Organize abstracts as follows:

1. Begin with a strong sentence that summarizes the entire proposal. The sentence should identify exactly what you want and why you want it.

2. Identify the problem you are researching or the market need.

3. Explain how your team will solve this problem or meet this market need.

Capturing all this information in a small space is a tall order. Every word counts. If necessary, reread Section 2 of this book for helpful hints on keeping it short. For example, consider the following abstract for a research proposal:

Abstract

We request a grant of $2.85 million to fabricate, install, and test 1,400 roof destressing systems in hurricane-prone coastal zones. In major hurricanes (category 3 or higher), 30% of all houses within two miles of the coast incur significant structural damage. Hurricane Andrew (1992) alone caused $30 billion in property damage, primarily to single-family structures. High winds cause huge pressure differentials between various parts of a roof, which can lead to catastrophic roof damage. Our team has developed a prototype for a new roof de-stressing system. This system uses aerodynamic principles to shunt high winds away from roofs. This system has proven highly effective in computer simulations and wind-tunnel tests and is now ready for real-world testing. We propose identifying 56 hurricane-prone coastal zones with new construction and placing 25 test devices in each zone.

In the preceding abstract, note the following:

- The opening sentence identifies the amount of the request ($2.85 million) and the reason. Reviewers who only read the first sentence would probably have a decent idea what the proposal is about.

- The next three sentences demonstrate an understanding of the problem by stating several numerical facts.

- The concluding four sentences state a solution to the problem and offer some reassuring preliminary evidence (computer simulations and wind-tunnel tests). The final sentence summarizes the test methodology.

Proposal Element: Contingency Plans

By their nature, proposals are glass-half-full sorts of documents. No one writes a proposal intending to fail. However, unbridled optimism hardly matches reality. After all, very few research projects produce important breakthroughs, most new businesses fail after a few years, and publishers lose money on the majority of technical books they publish. Failure is not a secret—the people reviewing proposals are certainly aware of it—yet proposals rarely cover failure. This practice is almost superstitious, as if mentioning failure will lead to it.

In certain kinds of proposals, you should describe possible failures and explain how you will handle them. For example, consider an excerpt from a research proposal on a new pharmaceutical. The following proposal should please members of the proposal review committee because it will prevent funding 18 months of worthless experimentation:

> We expect phase 1 clinical testing to demonstrate significant reductions in long-term acne scarring. Although this testing will last 24 months, we expect to have valuable preliminary results after only 6 months. After 6 months, if Cream A is proving less than 70% as effective as expected, we will halt the clinical trial on Cream A and focus our resources on improving Cream B.

Now consider a portion of a business plan that describes an alternate pricing policy based on information that cannot be known in advance. Notice that the following passage describes the alternate pricing policy as a contingency rather than as a response to a possible failure:

> Since we are targeting Fortune 500 customers, we plan to sell the software exclusively through flat-fee site licenses. However, if our sales force uncovers significant opportunities in medium-sized companies, we will alter the pricing policy in year 2. If a market for medium-sized companies develops, we will offer per-server licensing on a subset version of the base product.

Describing failure is not appropriate for all proposals. In some situations, you only get one chance to do it right. Book publishers, for example, generally won't let you rewrite an unsuccessful book. Nevertheless, since books usually take at least a year to write, consider describing possible mid-course corrections should the technology change while the book is being written.

Proposals for Revolutionary Ideas

Scientists and engineers often dream up improvements to existing theories, technologies, and products. Proposals that explain how to improve an existing technology usually find a receptive audience. However, at some point in your career, you are bound to dream up something wonderfully new and revolutionary. Unfortunately, revolutionary ideas are usually the hardest to sell. In a world where we are exhorted to think outside the box, few people are actually rewarded for doing so.

From the perspective of your audience (proposal reviewers), revolutionary proposals are risky. In a business plan, reviewers worry about who would buy this revolutionary product and how it would be marketed. Therefore, an effective revolutionary proposal must focus on minimizing the appearance of risk. The following list offers a few suggestions:

- **Compare certain aspects of the idea with concepts that are stable and successful.** For example, although your product might be revolutionary, explain that your pricing model is just like Microsoft's.

- **Inoculate your proposal from reviewer skepticism.** Answer the questions that will naturally come to mind.

- **Portray yourself as a down-to-earth, well-organized, focused individual.** Don't portray yourself as a creative dreamer. Many reviewers fear that highly creative people don't have the organizational skills to bring the dream to reality.

- **Avoid describing the idea as revolutionary.** Instead, build your case and let the reviewer come to that conclusion.

For example, the following passage from a business plan uses the preceding techniques:

Technical Background (for a revolutionary idea)

Leveraging Nobel laureate Louis Tucker's risk theory, our team can predict commodity prices approximately two hours in advance. Many people have gone broke making this claim; however, we have three years of extensive data to back it up.

Our team consists of four PhD mathematicians and a commodities trader. We have methodically developed pricing formulas based on over 150 parameters. Our team has gradually refined the weightings of each of the parameters to yield accurate predictive formulas for three different commodities.

Research Proposals

Scientists in the academic world submit proposals to perform research. A research proposal states a problem and your solution to it. Good proposals succinctly state the problem and provide a more comprehensive solution. The competition to write effective proposals is scary since a good grant can make a career. If you are competing for grant money against well-endowed labs, be aware that most of them already have a dedicated, experienced proposal writer on staff.

Understanding the Audience

The people reviewing research proposals are almost always scientists in your field. (In some cases, an interdisciplinary team reviews proposals.) Each proposal review committee includes one leader.

Committees typically hire a project manager (often, a nonscientist), whose duties include interfacing with submitters. Feel free to ask the project manager for bios of the people who will review your proposal.

Strategy

I hate to start with a negative, but it is useful to consider why research proposals get rejected. Possible reasons for rejection include the following:

- The proposed research did not impress the committee; in other words, the committee just didn't think the results would be that valuable.

- The committee wasn't impressed with the proposer's credentials.

- The proposal was fuzzy. (If only the proposer had read, *Spring Into Technical Writing for Scientists and Engineers...*)

The following strategies should improve your odds of acceptance:

- **Excite your reviewers.** Make sure that your proposal conveys your own excitement about doing the research. The committee reads many dull proposals; make your proposal sparkle.

- **Ally yourself with someone who will impress the committee.** Ideally, science ought to be blind to this sort of thing, but let's face it—the review committee is much more likely to turn over money to someone with a track record than to an unknown.

- **Write clearly.** The writing in most proposals is generally rather cloudy. By submitting a well-written document that clearly defines your proposed research, you might just shock the review committee into giving you money. Just follow the rules identified in Section 2 of this book and you'll have a huge head start.

- **Write concisely.** For example, the National Science Foundation (NSF) requires that proposals be no longer than 15 pages.

- **Make a case for yourself.** Your knowledge of the literature and your experience should convince reviewers that you are the best person in the world to do this research.

Contents of a Research Proposal

Many review committees provide an extensive template; others provide extensive guidelines but expect you to figure out the details within those guidelines. As always, proposals previously accepted by your target committee are the best teachers, so read some.

Research proposals typically contain some subset of the following items:

- cover sheet or cover letter

- abstract or project summary

- table of contents

- project description, which should include at least the following:

  - significance statement

  - objectives and hypothesis

  - experimental design and methods

- bios

- schedule

- budget, which is often quite detailed and might contain elements of a business plan in order to justify the money

The project description is the heart of every research proposal. You have a few minutes to explain what you plan to do, why it is important, and how it compares to other research in your field. The next few chunks detail the key aspects of a project description.

Research Proposals: Significance Statements

The **significance statement** is a one- or two-page explanation of why your research is important. Don't explain why the research is important to *you*. Instead, figure out why the research would be important to the committee. Read the committee's mission statement or template carefully. Committees funded by government organizations might require that research proposals further some political need. Their idea of importance may have very little to do with your idea of pure science.

The following (fictitious) significance statement is aimed at a (fictitious) government agency that has previously funded projects to improve public safety.

One tactic for significance statements is to start general and gradually get more specific. For example, the following sample starts with a broad statement (hurricanes cause casualties), moves down to the reason (poor predictions caused by imperfect technology), and moves further down to the specifics of this project (our solution improves predictions).

Significance Statement (sample)

The great Galveston hurricane of 1900 caused approximately 10,000 deaths, and the Lake Okeechobee hurricane of 1928 caused over 1,800 deaths. Fortunately, as forecasting skills have improved, the number of casualties has dropped dramatically For example, Hurricane Andrew of 1992 caused fewer than 30 deaths despite being stronger than the 1900 and 1928 storms. Nevertheless, even a single casualty from a hurricane is too great.

A large part of modern casualties are caused by the "boy-who-cried-wolf" phenomenon. When forecasters issue an evacuation order and the storm misses, residents stop taking evacuation errors seriously. The next storm thus causes even more casualties. Superior predictions will eliminate this phenomenon.

The National Hurricane Center (NHC) currently bases predictions on five diagnostic programs (NOGAPS, GFS, BAM, UKMET, and GFDL). All of these programs average at least 100 miles in errorsa over a 72-hour forecast period and over 300 miles in errors over a 120-hour forecast period. These programs are 78% better[b] than the previous generation of forecasting tools; however, they still have relatively wide margins of error.

We have developed a prototype for a program that will cut forecast errors in half within two years. When working with preliminary data, our current prototype already yields hindcast accuracy 20% better than current NHC models.

a. citation [*Omitted for space reasons.*]
b. citation [*Omitted for space reasons.*]

Research Proposals: Objectives and Hypotheses

The project description should include a statement of your project's objectives. In other words, if all goes well with the project, what do you hope to produce? For example, the following sample "Objectives" section concisely identifies the desired results.

Objectives (sample)

We aim to develop a hurricane forecasting program that will cut forecast errors to the levels shown in Table 1.

TABLE 1 Mean Forecasting Errors: A comparison

	72-Hour Mean Error (in miles)	120-Hour Mean Error (in miles)
Current Forecasting Programs	100	300
Our Forecasting Program in Two Years	50	125
Our Forecasting Program in Four Years	25	85

Your objectives should be crystal clear and contain no logical fallacies. (Can you spot the subtle fallacy[1] in the preceding "Objectives" section?)

The project description should also contain one or more hypotheses. As with any research, your hypotheses should be provable. Unless the template permits a tremendous amount of detail (which is unlikely), you should describe only one or two primary hypotheses and omit any secondary or tertiary hypotheses. For example, consider the following:

Hypothesis (sample)

Most current programs forecast motion by calculating the fluid forces that the surrounding environment exerts on the hurricane and weighting the results with climatology. We hypothesize that it is more accurate to study not only the influence of the environment on the hurricane but also the influence of the hurricane on the surrounding environment. Our secondary hypothesis is that climatology can be removed completely from the forecasting program.

---

1. The table compares the behavior of the *current* models to the proposer's *future* models. The performance of the current models is also likely to improve over the next two to four years.

Research Proposals: Design and Methods

The "Design and Methods" section of a research proposal summarizes the experiments that you plan to perform. Again, most proposal templates don't permit you much space (a page or two), so you must summarize instead of elaborate. A typical organization for this section is as follows:

- **Overview.** Provide a good introductory paragraph about your experiment.

- **Methods and Materials.** Explain your technique and the experimental tools you'll need. Where appropriate, identify any techniques pioneered by others. If you are proposing a multipart experiment, summarize each of the major parts.

- **Data Analysis (or Evaluating Results).** Describe the statistical tests your team will apply to the data. Explain why your chosen method of analysis is appropriate.

For example, consider the following experimental "Design and Methods" section:

Design and Methods (sample)

Our algorithm breaks down the interaction of hurricane and environment into a series of 15-minute "steps." After each step, our program examines deltas in both the hurricane's three-dimensional shape and in the surrounding environment. The algorithm models this pas de deux, readjusting the dance floor after each move.

Methods and Materials. Our team will evaluate environmental data provided by the National Hurricane Center (NHC) from 1992 to present. (Data sets prior to 1992 do not contain enough detail for our purposes.) These data sets are very large and require a tremendous number of floating-point calculations ($\sim10^{15}$) to compute a single 120-hour forecast. To handle this load, we require an 8-CPU server with 8 GB of RAM.

We plan to write approximately 1.2 million lines of C code and to take advantage of approximately 3.5 million lines of existing public-domain C code.

Data Analysis. The NHC currently uses a simple standard metric for determining the accuracy of predictions. Every six hours, they measure the distance between the actual and predicted positions of the storm. The mean of these measurements yields the accuracy of the forecast program. To ensure meaningful comparisons, we will use this standard accuracy metric as well.

As Zeven (2001) notes, seasons in which storms move slowly yield greater projective accuracy, even without any actual improvement in diagnostic programs. Therefore, we will also supply a second metric that factors in the speed of each storm.

Book Proposals

Writing a book is one of the best things that you can do to improve your career.

> **The Opening Moves in Writing a Book**
>
> A nattily dressed publisher at a conference caught your presentation and asked if you'd consider expanding the topic into a commercial book. You were hoping he would ask, so you launched into a sparkling 60-second pitch on the book you've always dreamed of writing. He liked it, so now is your chance to develop that 60-second pitch into a book proposal.

Understanding the Audience

Publishers are erudite, highly intelligent, well-spoken, friendly people who are typically only interested in one thing:

> Selling lots of books

Publishers know every mover and shaker in your field and will drop names to prove it. Publishers also know every book published in your field. They know which of these books sold well and which did not.

Some technical publishers have a strong background in your field. Some studied the field in college or graduate school, and a few may even have worked in the field. Technical publishers typically attend a lot conferences and are up to date with current trends and buzzwords. However, full-time publishers are *not* current practitioners of your field and won't know nearly as many specifics as you do. Publishers run wide but not deep.

Most people believe that authors come to publishers with ideas. Indeed, that does happen occasionally, but usually publishers think up titles and then go in search of authors. If a publisher approaches you for a book, then chances are decent that he's already figured out that the book will sell.

Strategy

Publishers' pay is tied to how many books they sell. Those publishers that sell only a few books will soon be searching for another job. (Publishing jobs are highly competitive, and turnover is astounding.) Therefore, the overall strategy is "simply" to convince the publisher that you can write a book that will sell lots of copies. To do so, your book proposal should convey the following:

- **You can write well.** The best way to prove this is to have already written a successful book. If you can't make that claim, a *well-written* article in a magazine will help boost your cause. The ultimate proof of your writing skill is submitting a beautifully clear, well-reasoned book proposal that follows the publisher's template. (Yes, the pressure is on.)

- **You understand the market.** Your proposal must explain who the competitors are and why your book will beat them. Remember—publishing is all about money.

- **You can meet the schedule.** Publishers love writers who hit all their deadlines. Writers who miss deadlines cost the publisher money.

What Does Not Impress Commercial Publishers?

The following will *not* improve your chances with publishers:

- **Fame.** Unlike fiction, consumers rarely buy commercial technical books based on the author's good name. (On the other hand, if you are producing a book for an academic audience, intellectual fame helps considerably.)

- **Lab reports.** Publishing many reports in research journals proves that you are an important researcher but doesn't prove that you can write a good book.

Contents of a Book Proposal

Most publishers provide extensive book proposal templates, which include a subset of the following items:

- abstract
- biography
- list of previous publications
- marketing
- detailed outline
- schedule

The following page contains the marketing section of a proposal for a book on a fictitious new programming language named *Fenster*.

Book Proposal: Example Marketing Section

Who are the primary and secondary audiences for your book?

I will aim this book primarily at professional programmers who have never programmed in Fenster before. Such programmers typically already own many programming books. Members of this audience may already own one of the competitive titles, but they prefer to buy at least two commercial manuals when attacking a new programming language.

The secondary audience consists of students. Currently, Fenster is very popular with computer science and traditional science students. However, because of financial constraints, this audience rarely buys commercial books unless a professor requires the books for class. Fortunately, several universities have just begun to offer Fenster courses.

What are the competitive books? How will your book compare to them? What is unique about your book?

The competitors are as follows:

- *The Joy of Fenster* (by Arnold Ziff) is the most popular book on this topic, primarily because it was the *only* book on the topic for almost a year. Reviewers on Amazon were annoyed by the lack of examples in this book. My book will contain more than twice as many examples as this one.

- *Fenster Goes to Monte Carlo* (by Leonard Hacker) is the second most popular Fenster title. The book features a lot of humor (as will mine), which seems to appeal to the Fenster community. The book's coverage of window manipulation (a crucial feature of Fenster) is very weak; my book will provide extensive coverage of window manipulation.

What is the appropriate price for your book?

Both of the competitors sell for a list price of $34.95. I propose to list the book for $29.95 to compete partially on price.

What are the proper venues (e.g., commercial bookstores, academic bookstores, conferences, and so on) to sell your book?

Both of the competitors are selling briskly through online and traditional bookstores, so these should be the primary venues. Since Fenster is catching on at universities, we should aim a secondary marketing effort at academic bookstores, particularly those selling to computer science departments.

The annual Fenster User Group (FUG) conference would be a natural place to sell this book. Leonard Hacker made a presentation at FUG last year, and his publisher was present to take orders.

Business Plans

Business plans aim to get money to start a new business or to expand an existing business.

Understanding the Audience

The target audience for business plans includes potential investors, often venture capitalists. Members of this audience read business plans to determine whether they can make a healthy return on their investment.

Some venture capitalists started off as technologists and gravitated towards business. Other venture capitalists have no formal training in technology but are conversant with it. Your business plan should assume a technologically sophisticated audience—one that is reasonably comfortable with jargon but has very little interest in deep technical detail.

Venture capitalists typically know your market extremely well. They will know which companies have succeeded and which have failed. They will have deep-seated theories on what works and what doesn't.

Strategy

Many technologists mistakenly believe that business plans should focus on technology. True, a business plan must explain the business's underlying technology; however, it must focus on how the company will make money and reward its investors.

A business plan is essentially a request to form a financial partnership, oftentimes with a stranger. Under what circumstances would you invest money with a stranger? The following list might be helpful:

- **The stranger has business experience.** Investors are understandably skeptical of business novices.

- **The stranger's business is already successful.** Note that existing companies that want to expand often write business plans.

- **The stranger explicitly states how investors will recoup their investment.** Investors want to cash out after a few years, so a good business plan explains how the investors will transition out of their investment.

- **The stranger's ideas are clear and sensible.** The proposal doesn't hide anything.

- **The stranger has identified a market and a plan for selling to this market.** The proposal details the size of the market and a plan to beat the competitors. Furthermore, the proposal handles contingencies for a shifting marketplace.

- **The stranger is no longer a stranger.** Again, would you really want to invest money with someone you didn't know? Probably not. Once you identify a group that invests in your kind of technology, try to find a connection with one of the investment principals. Maybe a friend of a friend could help you.

Contents of a Business Plan

Business plans are less "templated" than other kinds of proposals, so you have a fair amount of organizational freedom. Nevertheless, a good business plan contains at least the sections listed in Table 13-1.

TABLE 13-1   Sections in a Typical Business Plan

Section	Details
Front matter	Provide a cover letter (see "Proposal Element: Cover Letters" on page 184), cover page, table of contents, and so on.
Executive Summary	Provide a quick summary or abstract for busy readers. (See "Proposal Element: Abstracts" on page 187.)
Company	Provide a brief corporate history. Detail how the company is currently funded and who owns the company. Provide bios for your company's management team. (See "Proposal Element: Biographies" on page 185 for help on bios.)
Product	Explain the technological basis for the company. In other words, define what you are trying to sell.
Support	Explain who will provide customer support.
Market	Explain who will buy this good or service. If you are breaking into an existing market, describe the size of the current and future market. If this is a new market, provide realistic estimates, preferably based on reputable market research. What market share percentage can your company achieve? Be realistic.
Marketing	Describe your marketing plan. Explain how the proposed company will fulfill the market better than its competition. How big will your sales force be?
Manufacturing and Distribution	(If you are selling a service rather than a product, skip this section.) How will you manufacture your product? In what venues will customers buy your product? Do you have a distributor in place already?
Finances	Detail both your current financial position and your projections. What will you do with the investment? How will investors cash out?
Schedule	Detail what your proposed company will achieve and by when.

Summary of Proposals

Before submitting your proposal—while lying awake at 3:00 AM, wondering if a career in dry cleaning really would have been more satisfying—turn on the light and ask yourself the following questions about your proposal:

- Has someone edited the proposal for spelling and grammatical errors? (Do you really want to submit a proposal that contains typos?) Is the writing clear? Are the sentences a struggle to read? Are the fonts obtrusive?

- Has someone outside your team reviewed the proposal? The ideal reviewer is someone with similar credentials to the proposal review committee's members. Does the proposal make sense to the reviewer? Can the reviewer find any logical fallacies?

- Has someone outside your team reviewed the abstract? E-mail only the abstract of your proposal to several colleagues and ask them to describe what the proposal is really about. If your colleagues don't get it right, consider rewriting the abstract.

- Have you followed the proposal template very closely? Have you looked at successful proposals and mimicked their style?

- Does your cover letter contain your current contact information? You don't want the committee to send the check to the wrong address.

- Does your bio do you justice? Will proposal reviewers be impressed by the person described in the bio? Will your experience impress the target audience? If not, have you allied yourself with someone who *will* impress proposal reviewers?

- If you are writing a business plan or a book proposal, does your business focus come across, or do you sound like a pure techie? Does your proposal provide a cogent description of the current market?

- Do your budget line items sound realistic? (Can you really buy two new computers for only $50?) Do your budget columns add up?

- Do you come across as a grounded, well-organized individual, or do you sound completely scattershot?

- Does your proposal handle contingencies effectively? Have you explained anything other than the best-case scenario?

CHAPTER 14

Internal Planning Documents

The previous chapter looked at documents aimed at readers outside your immediate organization. In this chapter, we turn our gaze inward to study documents that you write for people within your organization. Instead of the wolf outside your door, you are now trying to please the jackals in the next cube.

This chapter focuses on the following three kinds of internal planning documents:

- **Business proposals.** These recommend new products or technologies to the upper management of your organization.

- **High-level technical specs.** These summarize a new product or technology so simply that even the vice president of marketing can understand them.

- **Low-level technical specs.** These detail exactly how your own engineering team will build new products or technologies.

The key to all three types of planning documents is understanding who your audience is and what it needs to get out of each type of document. Dr. Chekirnov, the barefoot biologist in the next cube over, needs very different information than Sam Minyon, the well-shod gentleman who heads procurement in the corner office.

They Will Know You by Your Writing

Internal planning documents, particularly business proposals and high-level technical specs, are very important to your career. Writing a clear, well-considered proposal or spec will speak volumes about your organizational abilities. Conversely, cloudy specs suggest cloudy thinking.

Business Proposals

Business plans sell a new company to potential investors. **Business proposals** sell a new idea to an existing company. If you are an engineer with a great idea for something your company ought to be doing, then you would typically write a business proposal to sell your idea to upper management. Writing a business proposal that changes the direction of your company is invaluable to your career, particularly if the project is a success.

Understanding Your Audience

Aim your business proposal at the upper management in your company. Note that upper management includes not only engineering management but also management in completely nontechnical areas. Further note that the top management in an engineering organization is typically years removed from hands-on use of technology and that many members of upper management no longer understand a lot of the nitty-gritty technical issues that you take for granted.

What Motivates High-Level Managers?

Upper management aims to enhance income and reduce expenses by making sound business decisions. That's the standard answer, anyway. The real-world answer is far more complex. For example, some top-level managers see their purview as a fiefdom to be zealously protected, particularly against all forms of change. For this reason, even the most wonderful business proposals sometimes get rejected without due consideration.

The clever writer knows that a great business proposal can only advance as far as his or her political leverage allows. You should campaign for a business proposal by walking it around to upper-level managers and encouraging feedback. Once upper-level managers feel that they have a hand in the plan, the plan has a far greater chance of success.

The Keys to Successful Business Proposals

Effective business proposals persuade readers of the following points:

- The idea is a sound business decision; it will generate a profit or reduce costs.

- The project is achievable; it is not the fantasy of an idealistic dreamer.

- The idea will not hurt current sources of revenue; it may even enhance current sales.

To persuade upper management of the preceding points, use the following tactics:

- **Rely on market research.** Does the market you are proposing already exist? How much revenue does this market currently generate? How much will this market grow over the next few years?

- **Compare your idea to successful similar projects at your own or peer companies.** This is not to say that upper management only accepts proven ideas, but it is easier to sell an exclamation point than a question mark.

- **Explain how your idea will outsell your competitors' products.** Note that the first company to market is not always the most successful in that market.

- **Find a corporate champion for the project.** If influential people in the company believe that your idea is sound, your idea has a far greater chance of success.

Analysis of the Sample Business Proposal

A two-page sample business proposal begins on the next page. It describes a new, high-end exercise machine. Note the following points about this business proposal:

- This business proposal focuses on business principles, not technical details.

- This business proposal invokes the names of two corporate champions (Elena and Bob) primarily to announce to other readers that two members of upper management have already approved the idea. (Before using this tactic, make sure that other top managers don't hate Elena and Bob.)

- This business proposal contains very few adjectives and adverbs; instead, it offers facts and invites readers to come to their own conclusions about the barrels of money that competitors are making.

- This business proposal uses simple, numerically based graphics to make its points about competitors. Even people who don't read every word of proposals will probably still glance at the two graphics and get something from them.

- The graphics omit the sorry state of the engineer's own company: the fact that the company's own revenues have held flat is buried underneath one of the graphics. (The wise writer doesn't offend the vice president of sales when trying to push a new idea that will require her approval.)

- The purpose of the "Background" section is not really to explain virtual reality (VR) but to suggest that VR is a firmly entrenched technology. (Notice the use of the word *old*.) The writer did not want readers to think that VR was a new and risky technology.

- Figure 14-2, on page 205, inoculates against the cannibalization-of-market question by demonstrating that overall revenue will probably increase.

Business Proposal: Example

Our team recommends that we design, manufacture, and market a new virtual reality (VR) stair-stepper model. This business proposal summarizes our recommendations.

Synopsis

Our R&D team recommends designing a new stair-stepper machine with VR features. This machine, which we are dubbing *King Kong*, features VR goggles. Customers wearing these goggles will feel as if they are climbing up the outside of the Empire State Building as they stair-step. The new machine uses the same chassis as our existing 3500 model but layers on additional digital components to enable VR features.

Background

Virtual reality is a term several decades old that means portraying fantasy in as realistic a fashion as possible. The ultimate VR system would generate perceptions indistinguishable from reality. Commercially successful VR entertainment is now available at movie theatres, theme parks, and game arcades. Due to gradual drops in digital component prices, it is now commercially practical to produce exercise equipment that offers a powerful VR experience. We can use VR to wed exercise to entertainment.

The Current Market in VR Exercise Equipment

Two of our competitors—Calispindex and Pravda Mills—have already manufactured and sold exercise equipment with some VR features. These competitive models are as follows:

- **Calispindex SpinCycle VR20.** This stationary bike has a small flat-panel screen that projects images of a rolling rural countryside while the customer exercises. As the terrain changes, the bike reacts accordingly. For example, when the screen shows an uphill portion of the course, the bike becomes harder to pedal.

- **Pravda Mills Tread-1000.** This treadmill has similar digital components to the Spin-Cycle VR. While running on the treadmill, customers view actual images from the Boston Marathon course.

Both companies featured these two VR products in their last two annual reports. The pie charts in Figure 14-1, taken from data in their annual reports, illustrate the growing importance of VR equipment to these two companies.

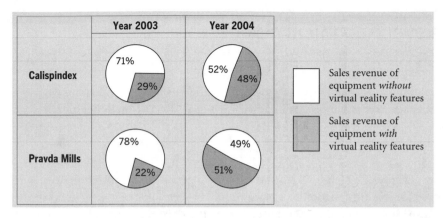

FIGURE 14-1 VR-based equipment accounts for an increasing percentage of sales.

Did the growth in VR equipment merely cannibalize the existing market in older, non-VR machines? In fact, as Figure 14-2 shows, both Pravda Mills and Calispindex enjoyed significant overall growth in new product revenue.

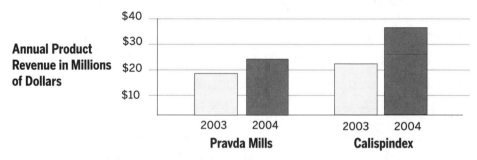

FIGURE 14-2 Overall sales at companies with VR equipment have grown.

During this same period, our own annual product revenue was flat.

Our Recommendation

When designing King Kong, we worked closely with Elena, vice president of marketing. She emphasized the need to enter this market as soon as possible. The components and manufacturing for this machine will cost us $750 more per unit than the 3500; however, we can sell the machine for $2,500 more per unit. Elena estimates that we can sell approximately 2,000 units in the first year and 6,000 units in the second year.

We estimate approximately nine months from project inception to first customer ship. Bob, vice president of manufacturing, feels that we can build this model in our existing Beloit factory.

High-Level Technical Specs

Once upper-level management gives the go-ahead to the business plan, the lead engineer or product manager writes a **high-level technical spec**, also called a **functional spec** in some industries.

Purpose

The primary purpose of a high-level technical spec is to get all the organizations in a company prepared for the introduction of a new product or technology. A really good spec helps unite disparate organizations; a really bad spec offends some of the department managers and ensures a bumpy ride.

Understanding Your Audience

This is the spec that launches a dozen specs. Managers read this spec in order to write specs of their own. Therefore, your audience skims over the spec, selectively attending to specific parts that will help them write their own spec. The purpose of your high-level technical spec is not to write the other specs but to provide enough information so that other managers can do their job. For example, your spec should contain information such as the following to enable a manufacturing manager to write the manufacturing plan:

- the parts required to build the new product

- a diagram or description suggesting how the parts fit together

Remember that your audience includes many people with little or no scientific or engineering background. The vice president of sales understands how to sell products, not how their innards work.

QUANTUM LEAP
Effective high-level technical specs must walk a thin line. The spec must provide information for other department heads without sounding as if you've made decisions that they are responsible for making. Your spec should specify which decisions you are expecting others to make.

Sections in a High-Level Technical Spec

The sections in a high-level technical spec depend on the industry. The sections for specs in the pharmaceutical industry will be quite different from those for the software industry. The following section list, for example, would be appropriate for an industry manufacturing a good old-fashioned tangible good:

- Synopsis
- Components
- Assembly
- Schedule
- List of Issues

The synopsis is the most important part of the entire functional spec. Given the harried schedules of most executives and their general disinclination to read specs, many executives will only read the synopsis. For this reason, it is vital to create a concise, well-organized, and (dare I say it) *exciting* synopsis.

The technical overview should explain what other departments need to know about the product. Remember, you still have the low-level technical spec to detail everything your own engineering group needs to know about the product.

The schedule and the list of issues often serve as the basis for regular cross-functional team meetings. In the opening draft of the high-level technical spec, you typically don't want to give too many schedule details, particularly those for other departments. (Managers from other departments won't want you to get too involved with their schedules.) Nevertheless, you still need to put out some broad starting dates. Then, just sit back and wait for the whining to begin.

Analysis of the Sample High-Level Technical Spec

The sample high-level technical spec that begins on the following page straddles the skinny line between providing enough information and not making decisions for other departments. The following is a list of key points from that spec:

- The spec does not include information on pricing or marketing. True, this sort of information had to be in the business proposal, but now marketing and sales will do as they please in making the decisions.

- The spec includes a fairly detailed list of components, which will be useful for the procurement department. This list carefully distinguishes between components with specific suppliers and components with open-ended suppliers.

- The "Assembly" section contains a few critical details but provides openings that the relevant people in other departments must complete. In some high-level technical specs, details such as how the laptop will be mounted would already have been worked out, and the spec would contain complete diagrams.

- The schedule is fairly sparse. A project manager or project leader will ultimately generate a full schedule containing perhaps several hundred milestones for a project of this complexity.

High-Level Technical Spec Example

All department heads at Social Climber Inc. should read this functional spec. It highlights a new product that we will develop during the year. If all goes as expected, this new product will be the first of many virtual reality (VR) exercise machines that blur the line between exercise and video games.

Synopsis

This spec details the proposed King Kong Climber 4500, which is a high-end stair-stepper machine with a digital twist. As movie fans know, at the climax of the movie *King Kong*, the great ape climbs up the outside of the Empire State Building. The 4500 will let customers enjoy a similar experience (but without fighter planes shooting at them) while getting a workout.

The key to the 4500 is a pair of VR goggles. These goggles are essentially a very, very tiny computer monitor that displays images. As a customer climbs steps on the 4500, the goggles display images from the Empire State Building. For example, when the customer has climbed the equivalent of 100 feet, the goggles will display the view from 10 stories up at the real Empire State Building.

Technology Overview

All the mechanical aspects of the King Kong Climber 4500 are identical to those of the current Social Climber 3500. In other words, both rely on the same chassis, motor, and pedals. However, the 4500 has significantly more digital components than the 3500.

The DVD of images from the Empire State Building will probably come from a company named Acrophobics' Nightmare. Our legal department is in negotiations with them.

Note that the 4500 is the first of many proposed VR exercise systems. For example, we have recently proposed a 5500 model based on multiplayer virtual races up tall buildings. Please make sure that parts suppliers understand that we are in this market for the long run.

Components

In addition to the components in the 3500, the 4500 also requires the following:

- **A Zebra-5 kit, manufactured by Dexco Unlimited.** This kit has all the digital hardware we need preassembled. It is essentially a laptop without a keyboard or monitor. Dexco is Goggleplex's preferred OEM.

- **Goggleplex Omega 20, manufactured by Goggleplex Inc.** These are the virtual reality goggles that the Zebra-5 kit requires. There are no standards for VR goggles, so there is no alternative vendor for the goggles.

- **A CD containing the software to run the 4500.** R&D is creating the software.

- **A 15-amp power supply from any supplier.** Note that the 3500 model draws only 12 amps.

Assembly

The Zebra-5 provides its own chassis. We would like manufacturing to mount this chassis at the top of the Y-bar, using a mounting mechanism to be determined by the mechanical engineering and manufacturing departments. Note the following additional assembly requirements:

- A CD must accompany each kit.

- A hard-copy documentation set must accompany each kit.

Schedule

Table 1 contains a proposed high-level schedule for the 4500.

TABLE 1 Schedule for Initial Development and Deployment of the 4500

Department	Milestone	Date
All	Detailed plans due	Jan. 7
R&D	Prototype	Mar. 5
R&D	Software development functional freeze	Mar. 26
Marketing and QA	Beta test	Apr. 27–Jul. 27
R&D	Final software development freeze	Aug. 24
QA	Internal testing on final model	Aug. 31–Sept. 30
Documentation	Final manuals printed	Sept. 29
Manufacturing	Production of 1st unit for customer shipment	Oct. 6
Manufacturing	Production of 250th unit for customer shipment	Dec. 8

Preliminary Issues

The following issues must be tracked at regular team meetings:

- How do we document safety issues for climbing stairs while wearing VR goggles?

- Do we need to license the name "King Kong" for this product?

Low-Level Technical Specs

Low-level technical specs, called **design specs** in some industries, are detailed blueprints aimed at the engineers in a group who must implement some aspect of the new product or technology. The lead engineer or engineering manager writes such specs.

Purpose

Low-level technical specs explain the work to be done, who will do it, and when it must be finished.

Understanding Your Audience

Your audience consists of all the engineers in your group. Therefore, use plenty of jargon. For example, the spec should not waste time explaining the C programming language to engineers who already program in C.

A typical project consists of a set of subprojects, each implemented by different engineers. You are under no obligation to provide the same level of detail for each of these subprojects. For example, the spec should typically provide plenty of details for junior engineers but allow senior engineers more latitude. Nevertheless, to ensure a smooth project, the spec must clearly define the interfaces between the different subprojects. For example, the spec should define the data to be passed between two different programming modules.

Sections in a Low-Level Technical Spec

All low-level technical spec should contain a synopsis of the project and a detailed schedule. Beyond those two requirements, each low-level technical spec depends on the industry and the project.

Analysis of the Sample Low-Level Technical Spec

A sample low-level technical spec starts on the next page. Note the following key points from this sample:

- The "Synopsis" section aims to motivate the target audience. It relies on the prime engineering motivator, which is that success will lead to more interesting projects.

- The different "Team" sections are personalized to provide the appropriate level of detail for each team. For example, Randy is an inexperienced engineer, so the "Team A" part of the spec provides a lot of low-level details for him. Marietta and Eswar are highly experienced, highly independent engineers who flourish in a creative environment; therefore, the "Team C" part of the spec is quite open-ended and only lays out a few requirements.

Low-Level Technical Spec Example

The software engineering team (Randy, Jenny, Eswar, Jim, and Marietta) should read this spec, which details how we will implement the 4500.

Synopsis

As you already know, the 4500 adds virtual reality (VR) features to the 3500. To implement VR, the 4500 adds a laptop (minus keyboard and screen) that wirelessly transmits images to a pair of VR goggles. As a customer climbs, the images change with their virtual height. Our team must develop all the software to enable this marvelous act of digital deception.

This is our first VR product, but if we get it right, it won't be our last. If we can sell enough of these, the company will let us do even cooler products next year.

Hardware

The hardware implementation is much more complex than anything we've done in the past. Instead of using PROMs and a tiny bit of RAM, we're now working with all the hardware power of a laptop. All hardware components fit inside a standard laptop (minus keyboard and monitor), which will be mounted at the top of the Y-bar. Table 14-1 lists the laptop's components.

TABLE 14-1 Laptop Components

Component	Spec
CPU	Motonel 5 Series at 3.2 GHz
RAM	256 MB SIMMS running at 400 MHz
Disk drive	10 GB
DVD player	A 12X read-only player
Networking	Ethernet board + Goggleplex wireless card (signal is robust for ~2.5 m)

We'll be fighting over who gets to use the Goggleplex Omega 20 virtual reality goggles. The actual resolution of the Omega 20 goggles is 1024×800, but because the images are less than an inch away from the customer's eyes, the effective resolution is ludicrously good. We've already taken delivery of two units that our team can use for development and QA.

Software Requirements

Our group will develop the software for the 4500. However, we will buy the Empire State Building images from another firm. Note that we will rely on our usual source-code control system and build utilities. The OEM who assembles our laptop will preinstall Red Hat 4.3. Since all Goggleplex APIs are in C, we will write all code for this project in C.

I've divided the software project into three teams.

Team A (Randy)

Team A must create a small API to fetch the exerciser's workload every 0.5 sec and generate statistics from it. The statistics must include cumulative height climbed, calories burned, and total number of steps. Write the statistics into a 6000-element array. Open a socket, and listen for two possible requests:

- Request "1" means send the most recently updated row in the array.

- Request "2" means send the entire (nontrivial) part of the array.

Team B (Jenny and Jim)

Team B must rely on the existing Goggleplex display API to transmit images to the VR goggles. To gather the most recent height reached, use sockets to send Request "1." Then, map the most recent height to the appropriate image on the DVD. Note that successive images on the DVD show views 0.5 m apart. For example, image 10 shows the view from 5.0 m above the street, image 11 is the view from 5.5 m, and so on.

Team C (Marietta and Eswar)

Team C will develop the GUI. Note that the goggles are not only an output device but also an input device that can detect pupil motion (equivalent to mouse motion) and eye blinks (equivalent to mouse clicks).

The GUI should provide a quick shutdown to allow a panicked customer to exit immediately.

At the end of the session, the GUI should display a graph (based on data in the array) showing height versus time, which identifies the fastest and slowest parts of the session. To get the data, use sockets to send Request "2."

Detailed Schedule

[*Omitted for space reasons.*]

Summary of Internal Planning Documents

When reviewing a business proposal you have created, ask yourself the following questions:

- Did you hit the target audience? The target audience is supposed to be high-level managers and, possibly, the board of directors or various investors in the company.

- Is your business proposal too technical? It really shouldn't be very technical at all. Remember that most of the people reading it do not have a technical background.

- Is your business proposal persuasive? It should rely on cold, hard, numerical marketing information, leading readers to come to one inescapable conclusion: doing this project is the right thing for the company.

- Have you "presold" the business proposal by talking it over with various upper-level managers? Does the business proposal invoke the names of several important proponents?

When reviewing a high-level technical spec you have created, ask yourself the following questions:

- Did you hit the target audience? The target audience is supposed to be the people (often managers but possibly project leaders) who will write detailed specs for their own departments.

- Did you make the spec too technical? The head of procurement does not care what language you are coding in.

- Did you provide enough information for the head of procurement or head of manufacturing to write their specs?

- Did you provide too much information about marketing? You shouldn't provide much detail at all about marketing or sales.

When reviewing a low-level technical spec you have created, ask yourself the following questions:

- Did you hit the target audience? The target audience consists of all the people in your immediate engineering organization.

- Did you make the spec technical enough? This group really needs technical details.

- Did you provide enough information for the engineers to get started on the project? After all, that is the goal for this sort of spec.

CHAPTER 15

Lab Reports

A
lthough most scientists love doing research, many hate writing lab reports. This is a real pity because a muddy lab report can soil even the best experiments. Imagine researching a topic for years, getting valuable results, and then being refused for publication just because editors and reviewers did not understand what you are trying to say. Surely, though, that won't happen to you—not with your clear, concise, accurate writing style.

Lab reports typically consist of the following sections:

- Abstract
- Introduction
- Materials
- Procedure
- Results
- Discussion
- Conclusion
- References

The exact list of section names varies somewhat between disciplines and journals. For example, the "Materials" section is sometimes called "Methods and Materials" or "Equipment." Some journals require additional topics.

About the Experiment in This Chapter

The experiment described in this chapter never happened; I made it up. In addition, the citations and references in this chapter are also fictitious. The experiment is simply an example to illustrate how to write lab reports.

Abstract

The "Abstract" section presents the entire lab report in miniature. Abstracts must be quite short, running only a single paragraph. Journals impose different length and style requirements, to which you must pay careful attention.

Abstracts must be strong enough to stand on their own. Your abstract will probably be placed in a collection of abstracts. Readers browse these collections to determine which lab reports are worth reading. If you are trying to attract readers, you should expend a fair amount of energy on the abstract. Note that far more readers will read your abstract than will read the rest of your lab report.

QUANTUM LEAP

I recommend writing the abstract after you have written the rest of the lab report. By writing it last, you can lift key sentences from other parts of the lab report and use them (or slightly modified versions of them) in your abstract.

The following list presents a possible template for creating an abstract:

1. Begin with a sentence or two that summarizes the hypothesis that your experiment addresses. You may optionally phrase this opening as a rhetorical question.

2. In a sentence or two, summarize relevant research on the hypothesis.

3. In two or three sentences, describe the experiment.

4. In one to three sentences, summarize the results.

5. Conclude with a sentence explaining why the results are important.

Abstract

Does a child's focus correlate with barometric pressure? If so, does it correlate positively or negatively? Tucker (1999) hypothesized a negative correlation, but this assertion has never been tested. Our team used the MISHA CPT to measure the focus of a group of 150 third-grade students. We divided the students into three groups of 50 students. One group took the MISHA CPT when barometric pressure was low, another group took it when barometric pressure was neutral, and the final group took it when barometric pressure was high. The results found that children focused significantly better when barometric pressure was low than when barometric pressure was neutral or high. The results suggest that when diagnosing ADHD, practitioners should give the CPT when barometric pressure is neutral.

Introduction

Begin the "Introduction" section by identifying the purpose of the experiment. In other words, identify the question you are trying to answer. Try to identify the purpose in a paragraph or two, although for "wide" studies, you may need to provide multiple paragraphs or even multiple pages.

After stating the purpose, explain the theory (or theories) on which you have based the experiment. In your explanation, cite previous *relevant* research on this topic. (Don't digress too far into tangential studies.) Identify not only what previous researchers have done but also what their research did not cover. Citing research helps bring readers up to speed on the current thinking about this topic. In addition, citing research helps you avoid potential plagiarism charges.

As always, audience definition is essential—figure out what your target audience already knows and then supply the delta they must learn to understand your experiment. The audience for lab reports typically consists of highly educated experts, many of whom are fellow researchers. However, highly educated experts are typically highly specialized and might not be familiar with current research in the broader field.

The introduction should also explain any special equipment that might be unfamiliar to your audience. For example, consider a lab report based on an experiment in transuranium elements. If you are targeting the lab report at a journal that focuses on research into transuranium elements, your introduction should not waste the reader's time explaining the usual lab equipment in that field. If, however, you are targeting the lab report at a journal that prints all types of physics research, then your introduction should explain the apparatus.

The following sample introduction targets a periodical that publishes general psychological research for a broad range of psychologists. If the target periodical were devoted to the study of Attention Deficit Hyperactivity Disorder (ADHD), then the author would have omitted most of the background information about the Continuous Performance Test (CPT). Note that a real introduction might be considerably longer than the sample.

Introduction

This experiment seeks to determine whether barometric pressure influences a child's focus. In other words, does a change in barometric pressure change a child's ability to focus?

Tucker (1999) hypothesized that changes in barometric pressure might be responsible for the large variances in focus that the same child exhibits on different days. She suspected a negative correlation between barometric pressure and focus, Siska (2003) determined that focus changes within a child were negatively correlated with blood-sugar levels. However, Siska noted that the correlation was somewhat less than predicted and suggested that other environmental factors (including weather) might be responsible for the delta. Mackinson and Goldberg (2004) found that slight changes in ambient temperature accounted for relatively sharp changes in focus within the same child.

The literature does not provide any studies that correlate focus with barometric pressure.

Our experiment attempts to prove Tucker's hypothesis. We, too, suspect that children focus better on days when the barometric pressure is low than they do on days when the barometric pressure is high. Prior to beginning the study, we sent an informal e-mail survey to 50 elementary school teachers randomly selected in the Washington D.C. area. Of the 32 responses we received, 28 teachers indicated that students were "more manageable" on rainy days than on sunny days. (Barometric pressure is usually lower on rainy days than on sunny days.)

To measure focus, our team relied on the Continuous Performance Test (CPT). The CPT is currently the most widely used instrument for studying focus and for diagnosing Attention Deficit Hyperactivity Disorder (ADHD). The CPT is a computer program that looks to most children like a sort of primitive video game. The CPT displays a series of letters on a computer monitor. Children respond by typing certain keys on the computer's keyboard. The version of the CPT we used—MISHA CPT— lasted approximately 14 minutes. Although most clinicians use a CPT when diagnosing ADHD, the MISHA CPT is also an accurate assessment tool for defining any child's ability to focus.

Unlike temperature and relative humidity, barometric pressure is typically the same indoors and outdoors. Therefore, testing children indoors offers as much validity as testing them outdoors.

The MISHA CPT generates 12 separate parameters and a summation index. The summation index is an integer between 1 and 100, with 1 representing an almost infinite attention span and 100 representing severe ADHD.

Materials

The "Materials" section should state exactly what equipment you used, leaving practically nothing to the imagination. For example, the following list is not precise enough:

- light bulbs

- 24 rats

If another lab team wants to reproduce the previous experiment, will they know what equipment to get? Probably not. Furthermore, if your lab report generates controversy or suspicion, it would be embarrassing to admit that your lab report omitted key materials. A list like the following is more reproducible:

- 24 incandescent 60W light bulbs arranged in a 6×4 rectangular matrix (see Figure 2)

- 24 Wistar male rats, all between 10 and 12 weeks old

- a Hyperion Rat Habitat, Model 260-R

If your equipment falls into several categories, it is usually clearer to create several subheads and several distinct lists.

If the test subjects are humans, create a subhead called "Subjects." In this subsection, explain how you chose the subjects and any relevant information about their background.

Materials

Our team relied on the following hardware and software:

- 25 identical Dell laptops, each running MISHA CPT v2.1 on Windows XP

- A detached "desktop-style" Dell keyboard (not the laptop keyboard)

- A digital barometer: the Dexco Unlimited model B-16

Subjects

We tested 150 third-grade students chosen at random from a pool of 346 applicants from eight Washington D.C.–area public and private elementary schools. The students represented a fairly wide range of economic backgrounds. All agreed to participate in our study in exchange for a $25 gift certificate from a local toy store.

Experimental Procedure

The experimental procedure section should detail the *exact* experiment that your team ran. Do not idealize, do not fabricate, and do not aggrandize—just tell the truth. Describe your experiment so that another lab team could faithfully recreate your experiment.

If the experiment consisted of a series of steps that happened in a distinct order, present the procedure as a numbered list. Otherwise, present the procedure through standard paragraphs.

If your experimental setup is complex, nonintuitive, or just plain hard to describe, consider providing illustrations. A few good illustrations clarify an experiment and reduce potential ambiguity.

Experimental Procedure

We tested our subjects during the December school vacation week. To ensure a range of barometric pressure readings, we tested on three different days (Monday, Tuesday, and Thursday).

We randomly divided the 150 subjects into three groups of 50. Because of hardware limitations, we randomly divided each group of 50 into two subgroups (labeled A and B) of 25 subjects each. We tested subjects in Group A at 4:00 and Group B at 5:00.

Subjects took the test inside separate 4M^2 cubicles. Subjects could neither see nor hear other subjects during the test. The test consisted of the following steps:

1. Approximately three minutes prior to the test, experimenters ushered each subject into a separate cubicle in a large lab area.

2. The experimenters told subjects that the computer would provide instructions and assured the subjects that the test would only take about 15 minutes.

3. After a brief pause, the computer monitor displayed instructions for using the CPT while a recorded voice read the instructions aloud. (The text and audio instructions are the standard MISHA CPT instructions for children; see Appendix A for a transcript.)

4. At the conclusion of the test, an experimenter ushered each subject back to a waiting room, where subjects were reunited with parents or guardians.

The experimenters measured the barometric pressure at the midpoint of each test (approximately seven minutes after it began).

Results

The "Results" section details the quantitative outcome of the experiment, typically blending some combination of formulas, tables, and graphs. Keep the "Results" section objective and honest. (Later sections will give you a chance to be subjective.) Nevertheless, even while striving for truth and beauty, the statistics you choose to provide and those you choose to omit can easily bias your readers' opinions of the experiment.

Begin your "Results" section with a brief (one or two sentences) reminder of the overall purpose of the experiment. Follow that brief reminder with a brief description of the types of data your team collected. The prose within your "Results" section should introduce and summarize each graph or table. For example, the following paragraph helps focus the reader's attention on an important result:

> Figure 2 shows the absorption of I-131 as a function of the percentage of thyroid gland remaining. The graph is linear between 15% and 100%, but the slope changes markedly at around 10%.

If you have a lot of results, subdivide the "Results" section into multiple subsections. Do not give a subsection a meaningless name, such as the following:

Section A

Give each subsection a precise subtitle, as in the following example:

Absorption of I-131 by Percentage of Thyroid Remaining

Begin each subsection by identifying (in a short paragraph) what kinds of data this subsection covers.

Due to space limitations, the following page shows only a portion of a "Results" section.

Results

The experiment aimed to determine whether a child's focus correlated with barometric pressure. To that end, we examined the following types of data:

- barometric pressure
- changes in each child's focus over the three tests
- changes in the entire group's focus over the three tests

Barometric Pressure

Table 1 shows the barometric pressure during the tests. On Monday, barometric pressure was low. On Tuesday, barometric pressure was close to normal. On Thursday, barometric pressure was high. The barometric pressure varied very little between Group A and Group B within each day.

TABLE 1: Barometric Pressure on Test Days

Group	Monday (in Mbs.)	Tuesday (in Mbs.)	Thursday (in Mbs.)
A	1002	1015	1025
B	1002	1015	1025

Changes in Focus

Figure 1 shows three graphs representing the distribution of MISHA CPT summary index on the three testing days. (A lower summary index represents greater focus.)

[*Figure 1 omitted for space reasons*]

FIGURE 1: Distribution of summary index by day.

The subjects showed better focus on Monday (low barometric pressure) than on Tuesday or Thursday. The difference was statistically significant ($p=0.01$).

The subjects showed somewhat poorer focus on Thursday (high barometric pressure) than on Tuesday. However, the difference was not statistically significant.

The distribution for Thursday was much wider (std. dev. of 12) than for Monday or Tuesday (std. dev. of 6 and 8, respectively).

Discussion

The "Reports" section contains hard numerical data. In the "Discussion" section, you interpret that data, identifying themes and helping the reader draw conclusions.

Begin your "Discussion" section with a reminder of the hypothesis. Then, indicate one of the following opinions:

- The experiment's results prove (or recommend) the hypothesis.

- The experiment's results disprove the hypothesis.

- The experiment's results were inconclusive.

Build a case for your opinion. For example, explain why you think the results recommend the hypothesis. In building a case, answer questions such as the following:

- What results emphatically confirm your opinion?

- Is there reasonable doubt for your opinion? You are permitted the luxury of pointing out possible flaws in the experimental design or holes in the results.

- How do your results compare with similar experiments? Why might your results differ?

Your discussion can optionally provide alternate interpretations of the data or offer suggestions for further research.

Discussion

Our team attempted to determine whether barometric pressure influences children's ability to focus. In particular, we tested Tucker's (1999) hypothesis, which states that children's focus correlates negatively with barometric pressure.

The results show partial support for Tucker's hypothesis. In particular, children focus significantly better when the barometric pressure is low than they do when the barometric pressure is neutral or high. However, children focused only slightly worse during high pressure than normal pressure. The unusually high standard deviation on the high-pressure day (Thursday) suggests that high barometric pressure might affect some children greatly and others very little.

The results suggest that, when diagnosing ADHD, practitioners should give the CPT when barometric pressure is neutral.

The experiment covers only three different days. A more comprehensive experiment should sample at least 10 different days.

Conclusion

Many journals require a "Conclusion" section to summarize the results. At a minimum, the "Conclusion" section should provide reminders of the following information:

- the hypothesis
- the results

Some scientists present ideas for future research in the "Conclusion" section, while others place these ideas in the "Research" section. Some journals require that the conclusion is simply the final paragraph(s) of the "Discussion" section.

Conclusion

Our team tested Tucker's (1999) hypothesis, which states that children focus better when barometric pressure is relatively low than when barometric pressure is relatively high. Our results found strong evidence that children do focus better when barometric pressure is relatively low. However, our results did not find that children focus worse in high barometric pressure than they do in normal barometric pressure.

References

Lab reports must contain a "References" section that expands all the citations that have appeared in the report.

> **No Reference Format Is Truly Universal**
>
> Different disciplines impose different requirements for references. In addition, different journals within the same discipline sometimes impose different requirements.
>
> The information provided on this page is generic. To prevent antagonizing an editor, the best rule is simply to copy the reference style of previously published articles in the target journal.

In many disciplines, a "References" section consists of a numbered list. The numbers in the numbered list typically represent the order in which the citation appeared in the report. For example, the first citation in the report was for Tucker; therefore, element [1] in the "Reference" section is for Tucker.

> **References**
>
> [1] Tucker, M. J. 1999. "Focus and Weather," *ADHD Papers and Proceedings*, 42, 526–530.
>
> [2] Siska, N. S. 2003. "Temporal Focus Affect and Soft Drinks," *Psychology and Nutrition*, 77, 215–223.
>
> [3] Mackinson, J., and J. Goldberg. 2004. "Temperature and Focus," *ADHD Metrics*, 13, 482–491.

In other disciplines, the "References" section also consists of a numbered list, but the author presents list elements in alphabetical order.

Summary of Lab Reports

When reviewing your lab report, ask yourself the following questions:

- Who is the target audience for this lab report? Does your lab report hit the target audience?

- Would a person reading the abstract be tempted to read the remainder of the lab report? Would a person reading the abstract understand what your experiment was all about?

- Does your introduction provide sufficient background (or too much background) for your target audience? Does your introduction cover previous research relevant to this topic?

- Do your "Materials" and "Experimental Procedures" sections describe the experiment so precisely that other researchers could replicate the experiment exactly as you ran it?

- Does your "Results" section faithfully represent the data collected in the experiment? Have you checked the mathematical results carefully? Are the graphs and figures misleading?

- Does your "Discussion" section indicate whether the experiment verified or vilified the hypothesis?

- If someone only had time to read the "Conclusion" of your lab report, would he or she get some sense of what the experiment discovered?

- Does your "References" section cite all relevant research in the citation style that the target journal demands?

Grammar

Many journals require scientists to write lab reports in passive voice. If the journal requires passive voice, you must abide by this style. However, if the journal does not require passive voice, then you should write the lab report in active voice.

In nearly all types of writing, editors are sticklers about maintaining a consistent tense. For example, most technical manuals are written completely in present tense. Lab reports, however, are strange beasts. According to tradition, you mix tenses within a lab report. When describing the experiment (for example, in the "Experimental Procedures" section) or previous experiments, use the past tense. When describing theories (such as "Tucker's hypothesis"), use the present tense.

CHAPTER 16

PowerPoint Presentations

PowerPoint is the jacks-or-better of the corporate world—you've got to have it in order to stay in the game. Just try giving a seminar without PowerPoint or showing up at a meeting with, gasp, paper handouts. I live in mortal fear that my eulogy will be delivered as a broken PowerPoint stack. (*Damn it. Can't anyone get this projector to work?*)

PowerPoint gives the patina of professionalism to even the most amateur presentation. A few snappy bullet points, a stately background, and voilà—you've turned Fox into the BBC. PowerPoint presentations obfuscate facts, hide evil, and stifle questions. The devil is in the lack of detail. Everyone knows it, but everyone plays along anyway. Despite it all, you must master the art of PowerPoint presentations.

When asked to present some slides about your new invention to the vice-president of research and development, you might just be at a defining point in your career. Will you grab the funding, or will you get bogged down in a morass of mediocrity?

The Golden Rule of Presentations

Your ideas will fall flat if you cannot keep your audience's attention.

PowerPoint and the CEO

"We had 12.9 gigabytes of PowerPoint slides on our network. And I thought, What a huge waste of corporate productivity. So we banned it. And we've had three unbelievable record-breaking fiscal quarters since we banned PowerPoint. Now, I would argue that every company in the world, if it would just ban PowerPoint, would see their earnings skyrocket. Employees would stand around going, 'What do I do? Guess I've got to go to work.'" [Scott McNealy, CEO of Sun Microsystems, in an August 1997 interview with the *San Jose Mercury Times*.]

Organizing a Presentation: The Big Picture

Table 16-1 provides a rough algorithm for organizing a short PowerPoint presentation.

TABLE 16-1 Approximate Division of Time and Slides in a Short PowerPoint Presentation

Section	Approx. Percentage of Your Time and Slides
Introduction	5%
Body	75% to 80%
Conclusion	5%
Question-and-answer (Q-and-A) session	10% to 15%

For example, in a 20-minute presentation, divide your time as shown in Figure 16-1.

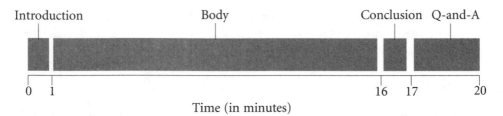

FIGURE 16-1 Time allocation in a typical 20-minute presentation.

If a presentation consumes more than 20 minutes, divide it into distinct units. Give each unit its own introduction, body, and conclusion. In addition, layer in a brief introduction and conclusion covering the whole presentation. For example, Figure 16-2 shows one way to organize a one-hour presentation into three units.

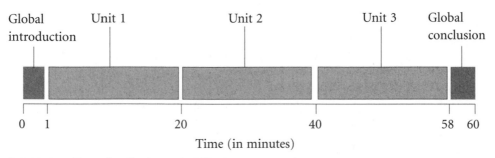

FIGURE 16-2 Time allocation in a typical 60-minute presentation.

The Number of Slides

Asking how many slides you need is kind of like asking a programmer the generic question, how many lines of code do you need to write a program? There just isn't enough information to answer this question precisely. Nevertheless, one popular rule of thumb states that you should estimate two minutes per slide. Personally, I find that audiences prefer more rapid presentations, so I like to estimate around 80 seconds per slide.

Parameters affecting the time per slide include the following:

- size of audience
- number of bullet points per slide and amount of text
- information density of graphics

As the size of your audience increases, you typically go through slides at a faster pace. Large audiences generally only interrupt your slides to ask questions during designated Q-and-A periods. Small audiences (seven or fewer) will feel much less self-conscious about interrupting you, and it may take several excruciating minutes to get through controversial slides. When audience members are highly familiar with each other, then the pace also slows. For example, if you are presenting to a 30-member lab team that has worked together for several years, you can expect plenty of interruptions. When audience members are primarily strangers to each other, interruptions drop.

Denser slides obviously take longer to consume, although sparse slides that require compensating voice-over also consume too much time.

Slides that contain complex graphics also take a long time to consume. They typically require long-winded voice-overs. So, for example, a presentation that contains mainly complex graphics might proceed at a pace of 5 or 10 minutes per slide.

The Opening Moments of a Presentation

I cannot underestimate the value of a positive first impression. In every communication medium—from movies to stand-up comedy to newspaper articles and right back to PowerPoint presentations—the first few seconds are the most critical. During those first few seconds, your audience will decide whether you are worth their focus. Succeed at the start, and your audience will root for you. Screw up the beginning, and you'll be scratching desperately for attention and approval.

So, how do you succeed in the opening moments? At the risk of sounding like a 3:00 AM infomercial, the key is confidence—you must believe in your heart of hearts that your presentation is valuable and that your ideas are critical. Dogs may or may not be able to sense fear, but audiences absolutely sense a lack of confidence. Humans attend to confident people; all the slick PowerPoint graphics won't bail the unconfident out of trouble.

Opening with a Joke

Some people tell jokes well. If you are one of those people, I recommend beginning your presentation with a relevant witticism. When you are one of many presenters, a little levity makes your presentation stand out from and rise above the others. Your presentation will be the one that people remember. Naturally, you have to sniff the mood of the room before telling a joke; in some situations, attempts at humor are destined to fall flat. Generally speaking, though, the joke is a little gift you present to your audience. Like most gifts, it is the thought that counts—the audience will appreciate you just for trying. In fact, the joke need not be particularly funny for you to succeed.

Always beta-test your jokes. Nothing ruins a presentation faster than offending someone in your audience. Make sure your beta testers include a wide range of people. One of the safer domains is to poke fun at your own profession.

By the way, spontaneous remarks always get a bigger laugh than prepared ones. So, prepare some spontaneous remarks.

Introductory Slides: The Traditional Approach

Since the opening moments are critical, your introductory slides better be good. The traditional approach to an introduction is to "tell 'em what you're going to tell 'em." I refer to this approach as the romantic-comedy style of overviews because your audience will know exactly how it is going to end just from watching the first few minutes. We will now consider a few examples of this style.

The slide in Figure 16-3 is direct and concise, but it is suboptimal. The simple title ("Introduction") does not mean anything. If someone were to come across this slide a few weeks after the presentation and read the title, would he or she remember what this presentation was about? What is this presentation really about, and why would an audience member want to pay attention to it?

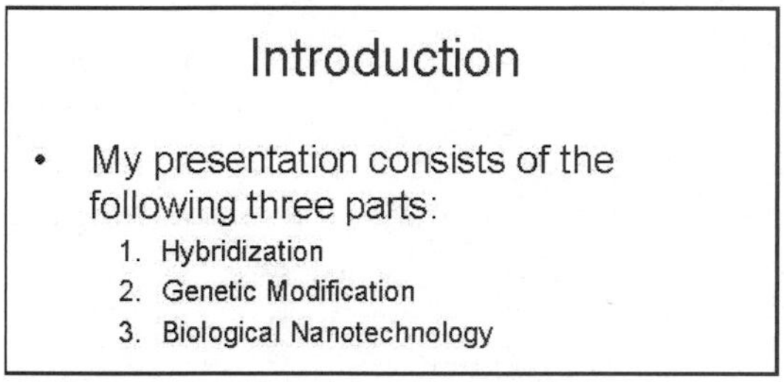

Introduction

- My presentation consists of the following three parts:
  1. Hybridization
  2. Genetic Modification
  3. Biological Nanotechnology

FIGURE 16-3 A suboptimal introductory slide.

The slide in Figure 16-4 offers a more instructive title. In addition, it contains date ranges, which technical audiences usually like and which are a nice device for building excitement. On the downside, the second bullet point ("where we've been") is a cliche. After reading this introductory slide, would you really know the true purpose of this presentation?

Creating Fruit Species

- Technologies to create new species of fruit:
  - 1900 to 2000: hybridization
  - 1995 to 2010: gene modification
  - 2010 to 2050: biological nanotechnology
- This presentation examines where we've been and where we're going.

FIGURE 16-4 A better title, but too cliche.

On the plus side, the slide in Figure 16-5 clearly states the purpose of the presentation (to provide an overview of creating new fruit species with biological nanotechnology). However, this slide looks like it was written by someone in marketing, which is the kiss of death for a presentation to technical people. To make the slide more appropriate for a technical audience, the writer should excise the word "exciting." The author of this slide is so overexcited that he provides almost exactly the same words in both the title and the first bullet, which is a waste of the audience's time. The final bullet will annoy all audience members who are proponents of current technologies.

An Exciting New Approach to Creating Fruit Species

- This presentation examines an exciting new way to create new fruit species:
  - Biological nanotechnology
- We'll also explore what's wrong with current technologies.

FIGURE 16-5 Too "exciting" for a technical audience.

The slide in Figure 16-6 has a solid title, a good explanation of the topic, and a straightforward approach that should appeal to a technical audience. The only negative is that it does not tell the order in which topics will be presented. To remedy this problem, the author must add a second slide that lists the order of presentation.

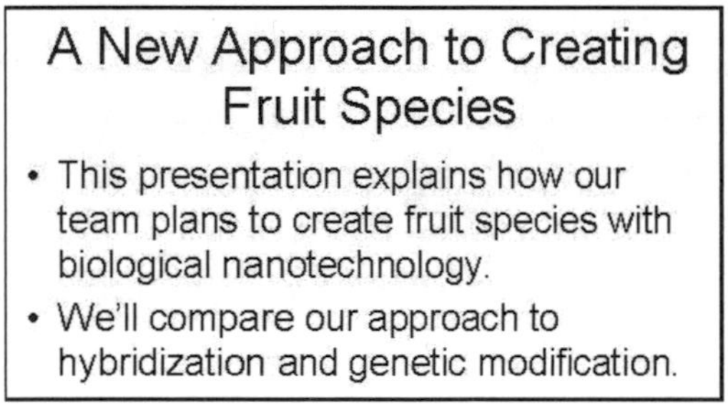

A New Approach to Creating Fruit Species

- This presentation explains how our team plans to create fruit species with biological nanotechnology.
- We'll compare our approach to hybridization and genetic modification.

FIGURE 16-6 Best of the bunch.

Introductory Slides: An Alternate Approach

An alternate approach for writing introductory slides is to "tease 'em what you're going to tell 'em." I refer to this approach as the mystery style of overviews because you withhold important pieces of the plot until just the right moment. You can build tremendous interest by promising a surprise and then layering in clues as you proceed. This technique often appeals to scientists since their life's work is essentially solving mysteries. Consider the gambit in Figure 16-7, which uses mystery style.

> ## A New Technology for Creating Fruit Species
> - Our team has a novel approach, which this presentation will uncover.
> - We'll compare our approach to hybridization and genetic modification.

FIGURE 16-7   Use mystery techniques to build suspense.

The title is clear enough to reassure audience members that they are in the right place. The first bullet announces the purpose without giving away the punch line. Humans are a naturally curious species, and intelligent humans are even more so. The audience wants to get to the next slide to learn the secret. Of course, the secret won't be found on the next slide. The secret must be nurtured while you gradually uncover clues. Occasionally, while laying out background information, you should remind the audience of the ultimate goal, as in the final bullet of Figure 16-8.

> ## Genetic Modification: Summary
> - Pros:
>   - Desired traits in a single generation
>   - More predictable than hybridization
> - Con:
>   - Expensive
> - Our new approach minimizes the expense.

FIGURE 16-8   Remind the audience that the secret will be revealed later.

In the perfect mystery presentation, you can slowly lead the audience to come to the same conclusions that you did.

Body Slides: Pace and Variety

Modern audiences have extremely short attention spans. Very few people can attend to a speech for longer than 20 minutes, and many attendees will be taking mental vacations after only 5 or 10. To audiences raised on television clickers, PowerPoint presentations get dull very quickly. When giving an hour-long presentation, how do you avoid just talking to yourself for the last 40 minutes?

To hold an audience through an hour-long presentation, you must switch focus every 15 minutes or so. After presenting a series of slides, get the audience to do something; for example, consider the following ideas:

- Ask them questions.

- Do an informal survey.

- Challenge the audience to solve a problem or to brainstorm ideas.

In general, keep the audience's attention cycling between the projection screen, you, and themselves. Never let the audience's eyes rest on one spot for too long; intelligent people go into a trance rather readily. Feel free to dim the projector occasionally and turn up the lights to force the audience's focus to change. Prepare for bland stretches by interspersing some interesting digression slides or examples.

Many people feel that a professional presentation requires that all slides have approximately the same length. In fact, this is not optimal. To keep your audience's attention, it is better to mix the length of slides; in other words, after a few long slides, insert a short slide.

When to Make a Presentation

Teachers and seminar leaders are well acquainted with "death valley," which strikes between 30 and 90 minutes after lunch. This is the worst time of day to give a presentation. Seated adults will sometimes fall asleep during this period. Similarly, avoid early morning presentations when speaking to teenagers.

The best time of day to present information is mid-morning (perhaps 10:00 AM to 11:00 AM) or two to three hours after lunch. On the other hand, if you are presenting bad news or speaking to an oppositional audience, schedule your presentation just after lunch, when people tend to be at their most relaxed.

Finally, do not rely too heavily on bulleted lists. Make sure that your slides mix in plenty of graphics, tables, and charts.

Mechanics: Fonts and Backgrounds

Many presenters obsess about mechanical points such as the appropriate font and backgrounds to use on slides. In fact, this is one of the classic procrastination techniques that many people use before buckling down and writing the copy for those blasted slides.

Don't waste your time worrying about fonts. PowerPoint provides excellent defaults for fonts and font sizes. If you feel a strong need to use a nondefault slide template, create one that uses a sans-serif font. Note, however, that the printed version of PowerPoint slides will look better with serif fonts. For more information on fonts, see Chapter 19.

Many companies and conferences mandate the background for PowerPoint presentations. If, however, the choice for backgrounds is yours, don't just pick a pretty one; pick one that is appropriate for your audience. If your audience is fairly stodgy and conservative, consider unintrusive designs with muted colors. If you are trying to appeal to a younger audience, find a more vibrant background.

When making a long presentation, consider color-coding different categories of slides. For example, you might paint a light green background on all slides containing exercises and a light blue background for all slides containing digressions.

For a multiunit presentation, consider giving each unit a distinct background image. For example, each slide in the unit on hybridization might contain a background image of a pea plant, while each slide in the genetic-modification unit might contain a background image of a microscope.

Body Slides: Effective Lists

Chapter 7 discusses lists. Beyond the general recommendations about lists in that chapter, consider the following additional suggestions for PowerPoint:

- Place rules before examples; in other words, define a rule and then supply examples to support the rule.

- Ensure that second-level bulleted items follow naturally from their first-level superiors.

- Minimize the number of list items at the third level.

- Keep each element fairly short.

The slide in Figure 16-9 violates most of these rules.

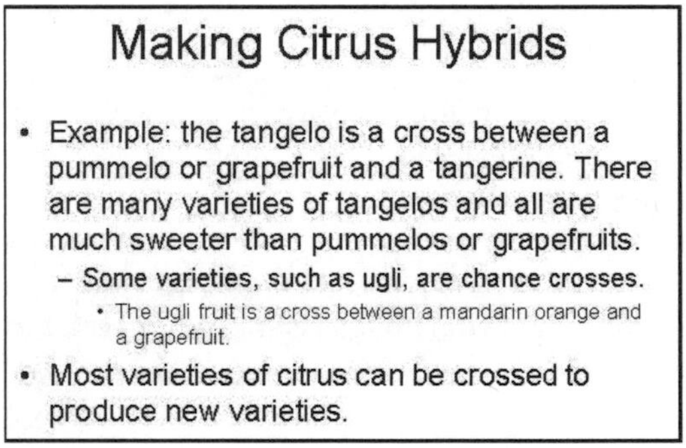

Making Citrus Hybrids

- Example: the tangelo is a cross between a pummelo or grapefruit and a tangerine. There are many varieties of tangelos and all are much sweeter than pummelos or grapefruits.
  - Some varieties, such as ugli, are chance crosses.
    - The ugli fruit is a cross between a mandarin orange and a grapefruit.
- Most varieties of citrus can be crossed to produce new varieties.

FIGURE 16-9 A slide with problematic lists.

The slide in Figure 16-9 contains some fine facts; however, these facts are defeated by the following faults:

- The title is misleading; it suggests that the slide will explain how to make citrus hybrids.

- The opening bullet is a borderline run-on, which makes the entire slide a bit too dense visually.

- The second-level bullet ("Some varieties...") does not follow logically from its superior bullet.

- The third-level bullet is an unnecessary digression.

- Examples are wonderful, but the general principle ("Most varieties of citrus can be crossed...") ought to precede the example rather than follow it.

The slide in Figure 16-10 is somewhat cleaner.

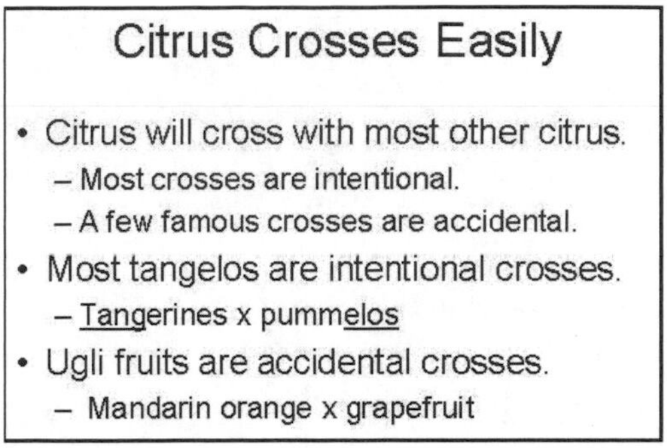

Citrus Crosses Easily

- Citrus will cross with most other citrus.
  - Most crosses are intentional.
  - A few famous crosses are accidental.
- Most tangelos are intentional crosses.
  - Tangerines x pummelos
- Ugli fruits are accidental crosses.
  - Mandarin orange x grapefruit

FIGURE 16-10   A cleaner set of lists.

The preceding slide focuses on the key points (citrus crosses through intention or accident) and provides supporting examples (tangelos and ugli fruits). The original slide mentioned the sweetness of the resulting fruit, but that fact was off topic, so the preceding slide omits it.

The slide in Figure 16-10 still contains too much information for some viewers. An alternate approach is to chop the single slide into two slides as follows:

- Place the rule (crosses are intentional or accidental) on one slide.

- Place the examples (tangelos and ugli fruits) on the next slide.

Yet another approach is to use a single short slide that focuses on the rule and then let your voice provide details of the examples. The slide might look like that shown in Figure 16-11. The downside to this approach is that audience members who look back on their slides won't have a record of the example details.

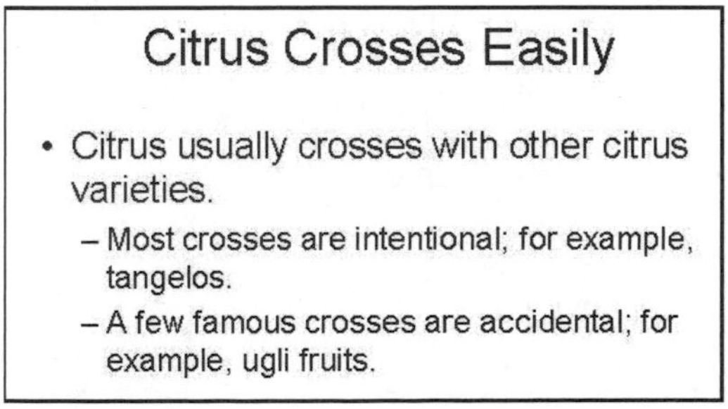

Citrus Crosses Easily

- Citrus usually crosses with other citrus varieties.
  - Most crosses are intentional; for example, tangelos.
  - A few famous crosses are accidental; for example, ugli fruits.

FIGURE 16-11   Focus on the rule in the slide; provide more examples in speech.

Audience: The Theory of Relativity

Your presentation's content and tone should depend on where you are in the organization relative to the people in your audience.

If the Audience Is Mainly above Your Level

Many presenters mistakenly believe that higher management has more technical knowledge than individual contributors. Although some high-level managers do retain their engineering skills, most high-level managers lose technical skills rapidly when they stop using them day to day.

High-level managers get paid to make critical decisions, typically based on financial considerations. Focus your presentation on money, answering questions such as the following:

- How much revenue will this provide?

- How much will this cost to produce?

- How much money will this save the organization?

From manufacturing to biotech to nonprofit, high-level manager ears typically only perk up in response to money and schedules.

If the Audience Is at Your Level

Individual contributors talking primarily to other individual contributors should focus on technical details. Slap on an extra coat of jargon, and keep the nerd-speak rolling. Ignore financial considerations; just get to the good stuff. Beware that this is the most likely audience to stumble down a rathole, so be prepared to shut off worthless debates dictatorially.

If the Audience Is below Your Level

If you are a manager presenting slides to your underlings, your audience may feel a certain sense of fright. They may wonder, *Why is she holding this meeting? Is my job okay?* Audience members are also nervous about being put on the spot. To get your audience to listen, it is sometimes smart to start the meeting with a perfunctory *all is well*.

Graphics

Most audiences view graphics as a treat. However, graphics are an expensive gift, taking long hours to prepare. If you have the time though, graphics do jazz up a PowerPoint presentation.

Although a truly gifted artist can tell a story entirely through graphic slides, the rest of us must learn to provide a mix of graphics and text. Mixing text and graphics on the same slide is okay if the following criteria are both met:

- The graphic is relatively simple.

- The text is brief and supports the graphic.

For example, the slide in Figure 16-12 meets both criteria.

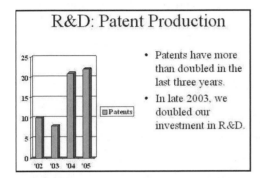

FIGURE 16-12   Simple graph and simple text shown on the same page.

If either the graphic or text is more complex than those shown Figure 16-12, place them on different slides.

Beware of Cliché Clip Art

PowerPoint provides a widely used set of clip art. A number of years ago, the novelty of clip art appealed to audiences. (*Wow—that clip art guy is shocked by our first quarter numbers, too.*) Today, though, the included clip art strikes most audiences as somewhat amateurish and cliche. All clip art is not bad. However, audiences now expect more than stock cartoon characters.

The Complexity of a Graphic

Before creating graphics, consider how your audience will view the PowerPoint presentation. Ask yourself the following questions:

- Will the presentation be projected on a high-quality screen, or will it be projected on a wall? Is the wall a bright color?

- Will your audience be close enough to view details in the graphics?

- Will you be able to point (with a light pen or stick) to selected details in the graphics while giving the presentation? (This is always handy.)

One of the classic mistakes in creating graphics for PowerPoint is providing too much detail, particularly when the viewing conditions make it almost impossible for your audience to notice detail. In such situations, frustrated speakers end up spending far too long explaining each graphics slide, which also frustrates the audience. Think it through: if viewing conditions are poor, reduce the detail in your graphics.

A reasonable formula for good graphics slides is as follows:

> One graphics slide should hold approximately the same informational content as one paragraph of text.

In other words, a graphics slide has about the right amount of detail if you need to speak about one paragraph's worth of words to explain it.

If a single graphics slide requires too much explanation, break it into two or more slides.

Printed Graphics in PowerPoint Presentations

Many presenters hand out hard copies of their slides to the audience so that their audience has something to write on when trading e-mail addresses and stock tips. Unfortunately, in hard copy, the graphics in PowerPoint slides are often hard to read. If hard-copy versions of slides are truly important, try to keep the graphics simple.

Question-and-Answer Sessions

In a successful presentation, your audience will ask lots of questions. (Lots of questions can also be a sign that you aren't explaining things very well, but we'll hope that this isn't the case.) Questions indicate interest. In addition, the rapid shift of focus between presenter and audience makes it harder for audience members to daydream. Two strategies for Q-and-A logistics are as follows:

- If you are speaking to a small audience, invite them to ask questions throughout the presentation. Encouraging questions helps keep your audience engaged.

- If you are speaking to a large audience, urge them to hold off on questions until the end of the presentation; however, if the presentation is lengthy, provide a Q-and-A session at the end of each unit.

The Q-and-A session frightens some presenters. After all, there is no script for it. In addition, questions can occasionally turn hostile. For best results, prepare for the Q-and-A session just as you would prepare for the rest of your presentation. Practice your presentation on a small group of colleagues, and invite them to ask questions. Have someone record these questions so that you can polish your answers later.

If a live practice session is not a possibility, you must imagine the questions that will be asked and prepare answers for them. Visualize your toughest critic (*That jerk!*) trying to pick your ideas apart. Practice your answers for the worst-case questions. You'll sleep better the night before.

Keep your answers short and focused—never ramble. Thank the first questioner for asking a question, as this will grease the wheels for subsequent ones. If you truly do not know the answer, say so. If you are taking an educated guess, say so.

Okay, you have done all your homework and are ready to take on all questioners and then... no one asks anything. This, too, can be rather uncomfortable. The wise presenter comes prepared with a list of dialog starters on an emergency slide. Turn the tables on your audience, and start asking them questions.

The final question of a Q-and-A session typically concludes the presentation. However, the final questions might wander off the mark or into minutia. Note that the first few moments and last few moments of any presentation are particularly important. Therefore, instead of ending with a final question, you should consider ending by displaying a final slide that truly wraps up your presentation.

Different Kinds of Learners

Just as most people have a dominant hand, most people also have a dominant learning style. In brief, most people are either visual learners or auditory learners. Visual learners acquire knowledge more naturally through what they read or see; auditory learners prefer to gather information through what they hear.

When giving a PowerPoint presentation to a group, both visual and auditory learners will be in your audience. The value of a PowerPoint presentation (as opposed to either a straight lecture or a book) is that you can simultaneously cater to both kinds of learners. What makes this even trickier is that the visual presentation proceeds at a much higher bandwidth than the auditory presentation; that is, people can read text much faster than you can speak it.

To cater to auditory learners, you should always provide a sound track; you may not simply throw slides up on the projector without providing any commentary. To cater to visual learners, you should always provide slides; you should not present new material orally without supporting slides. To complicate matters, a percentage of visual learners are distracted by auditory input. (Not everyone can read when music is on, particularly music with lyrics.)

Within the broad group of visual learners, some learn more readily from pictures than from text. Like a polite host offering a variety of dishes to suit many palates, try to provide a nice mixture of graphics and text.

When the Audience Is One

Your audience is not always a group. Sometimes, your audience is just a single person—often your supervisor. In this case, the wise presenter studies the target's learning style.

Does your target generally ignore your e-mail, preferring to get information from you in person? If so, your target is likely an auditory learner. Do not waste too much time writing elaborate slides; she won't read them anyway.

Does your target tend to ask you for written summaries following face-to-face meetings or insist that everything be written down? In this case, your target is likely a visual learner. A crisp executive summary page, followed by supporting pages of details will go a lot further than blah blah blah.

PowerPoint Speech: The Basics

Many technical presenters put far too much time into perfecting slides and not nearly enough time into practicing what they will say. The next few pages provides advice on improving the "sound track" to your PowerPoint presentation.

You must first perfect the following basics:

- **Speak loudly enough to be heard.** Project your voice. Imagine that you are launching your voice to the back row.

- **Speak clearly.** Enunciate the final syllable of each word—make sure you catch all the consonants.

- **Make eye contact with different people in different parts of the room.** If eye contact frightens you, stare at foreheads instead of eyes. (Your audience won't know the difference.)

- **Stand up straight.** Don't slouch. Square your shoulders. Bend your knees slightly. Breathe.

- **Drink something warm (not hot) just before speaking.** Never drink ice water.

When I Turn on the Projector, You Will Fall into a Deep and Satisfying Sleep

Your listeners, no matter how well educated, have a very short attention span. You are at war to keep their focus. To win the battle, avoid entrancing your audience. Just about any form of repetition can put intelligent audience members into a mild trance. To fight it, you must seek discontinuities by altering the following:

- **The speed of your delivery.** Slow down once in a while, and speed up once in a while.

- **The volume of your delivery.** Emphasize points by raising or lowering your voice.

- **The body motions of your delivery.** Never rock rhythmically. Never move your hands rhythmically. Feel free to walk around, particularly if you are using a wireless microphone.

If you must hold an audience's attention for a long time, try to change topics every 15 minutes or so.

PowerPoint Speech: Lessons from the Pros

You don't have to be a superstar performer to deliver a good PowerPoint speech. Nevertheless, technical speakers can benefit from the secrets of professional entertainers.[1] Treat your listeners as an audience whose attention you must hold. Learn to engage your audience. If you don't have their attention, your message will be lost.

These Tips Aren't for Every Audience

As with any form of communication, you must know your audience. If the forum at which you are speaking is very conservative, then beware of many of these suggestions. If, however, the forum is somewhat more casual, then take these suggestions seriously and emerge as the audience's favorite presenter.

Lesson 1: Entertainment Pros Provide a Pre-Show

A pre-show is, not surprisingly, the information that precedes the show. All forms of professional entertainment provide a pre-show to build an audience and to build excitement; for example, consider the following:

- Movies provide trailers.
- Theatres provide playbills.
- Television provides teasers and ads.

Technical speakers can sometimes provide a pre-show. Your pre-show might take the form of an e-mail to your audience. Perhaps you'll have someone distribute interesting handouts related to your presentation a few minutes before you speak. Perhaps these handouts will suggest something intriguing or puzzling.

Get the audience to anticipate your speech before you say your first word.

Lesson 2: Pros Deliver a Grand Finale

Good shows end with something big. Again, the beginning and ending of any presentation are the most memorable moments. Therefore, try to pull off a surprise just near the end, such as a fascinating revelation from your research. Better yet, promise your listeners something exciting but don't deliver on it until the end of the presentation.

---

1. I developed this chunk from my experience as a professional juggler.

Lesson 3: Pros Study Their Body Language

Many professional entertainers watch tapes of themselves. If you have never seen a tape of yourself giving a presentation, you should endure it soon. You will be amazed at what you discover. For added thrills, invite some friends to review the tape with you. Don't worry—this is just as "natural" a process as having a friend edit your writing.

While watching the tape, compare your body language with the following ideals:

- Stand up straight but not too stiffly.
- Lean forward slightly, which projects energy and enthusiasm.
- Project confidence by keeping your body expressions expansive and open. Never hunch your shoulders.
- Project optimism by moving your hands upward.
- Move around a little. Don't feel that you are tied to the podium. You can even move into the audience from time to time.
- Move your hands a bit. Just don't flap your hands enough to go into flight.

As noted earlier in this chapter, the beginning of any presentation is the most important part. Always stride confidently to the podium when it is your turn.

Some people in your audience are highly attuned to body language and will pick up all sorts of clues from it. Some will be trying to detect whether you are telling the truth, or at least whether you believe in what you are saying. Therefore, tell the truth and believe in what you are saying.

The Mute Button: Your Best Instructor

Would you like some help on body language? Consider watching television with the sound off. Watch your favorite comedians, and see how they invite you in through relaxed body gestures. Watch politicians try to convince you of their point of view through persuasive hand motions. Watch how the posture of news reporters conveys gravity or intensity.

Lesson 4: Pros Consider Costuming

Professional entertainers consider what to wear. They try to figure out what kind of costume will help their presentation.

Believe it or not, some engineers and scientists *do* attend to what you are wearing. Of course, most engineers and scientists take a certain pride in their fashion blindness. However, for those few in the audience that do care about your clothes, try to figure out which clothes will help your cause.

PowerPoint Speech: Overcoming Fear

If you are thumbing through this book in a bookstore, then there is a mighty good chance that you turned directly to this page. After all, a stunning percentage of people are afraid of public speaking. Climb a cliff? No problem. Explain to an audience how you climbed that cliff? Terrifying!

If the fear of public speaking is limiting your career, you should strongly consider working with a professional speech coach or possibly some sort of behavioral therapist (such as a hypnotist).

Many speakers fear the fear itself—they are afraid of freezing up and having a panic attack. They are also afraid that the audience will detect the fear and think less of them. These are certainly common fears, but you should know the following two facts:

- Even most professional performers experience stage fright.

- Experimentation strongly suggests that a certain amount of fear actually *helps* your presentation. This "fear" would seem to be our way of getting our bodies ready for a peak performance.

The best way to overcome fear is to practice the audio portion of your presentation over and over again. Practice it when you are in the shower. Practice it when you are in the car. Practice it right before you go to sleep. The social psychologist Robert Zajonc has noted that an audience helps the dominant response and hurts the nondominant response. In other words, if you *truly* know your speech, you will probably deliver it better in front of an audience than when alone. Conversely, delivering an unpracticed speech in front of an audience often leads to poor performance.

Practice Relaxing

You have undoubtedly heard someone tell you to relax by taking a deep breath. Everyone knows it works, but most people forget to breathe when tense. To force those soothing breaths, consider the following two techniques:

- If you will read from notes, write the stage direction "[deep breath]" at the end of every paragraph. In other words, script your breaths.

- If you will speak without notes, then when practicing, remember to practice taking a deep breath at the same points in your presentation. The deep breath will then become just as automatic as the rest of your speech.

Summary of PowerPoint Presentations

The night before giving your big PowerPoint presentation, obsess over the following questions:

- Have you rehearsed the entire presentation? Have you rehearsed multiple times?

- Is your opening strong? (Note that I'm saying *strong*, not *long*.) Are your first words interesting? Are you opening with a joke? Will that joke appeal to your audience?

- If your presentation is longer than 15 or 20 minutes, have you divided your presentation into discrete sections? Does each section contain a distinct beginning and end?

- Are your slides well organized? Does each slide follow naturally from the preceding slide? Are the slides organized into discrete sections?

- Are any of the bulleted list items run-ons? Do any slides contain too many list items?

- Do the text and graphics on each slide make sense?

- Have you fixed all spelling errors and typos?

- Is your presentation appropriate for your audience? Will supervisors or underlings take the presentation the right way?

- Do any of your graphics require more than a minute of explanation?

- Can your graphics be seen from the back of the room?

- Do you have a convincing and memorable conclusion?

- Will you be nervous? If so, do you have a scheme for fending off panic?

Finally, consider a few practical matters:

- Where is the printed material? Where are the notes? Where are the slides?

- Does the microphone work? Is it a clip-on microphone? If so, what part of your clothes will you clip it on to?

- Do you know how to work the projector?

- What will you wear?

Sleep tight.

CHAPTER 17

E-Mail

M any e-mail messages—even those sent by one brilliant engineer to another brilliant engineer—are unintelligible. This chapter focuses on writing effective e-mail messages.

Pickford Paradox of E-Mail

Mary Pickford—talented, intelligent, and charming—was a Hollywood superstar during the silent film era. After the advent of "talkies," Pickford noted that talkies were artistically inferior to silent movies. Therefore, in her opinion, silent films should have come after motion pictures with sound. Critics referred to her idea as the **Pickford Paradox**.

As a scientist or engineer, you might be thinking that the Pickford Paradox is ridiculous. After all, when would a lower-bandwidth medium (silent films) be superior to a higher-bandwidth medium (talkies)?

The strange answer is that we are living through a Pickford Paradox of our own right now. The bandwidth of a telephone call far outstrips that of an e-mail, and yet, corporate employees now communicate more through e-mail than through the telephone. Imagine if e-mail had been invented in the nineteenth-century and the telephone in the 1970s. If that had happened, we would probably still be amazed by the power of the telephone. (*Wow, you can have a conversation in real time with your voice!*)

Even more fascinating, the effective bandwidth of many e-mail messages is exactly zero. That's because many e-mail messages are so poorly written that no information actually gets communicated.

The Essence of the E-Mail Problem

Tom researched some software tools and then used Microsoft Word to compose the following short, formal document:

Recommendations for Performance-Analysis Tools

Sam asked me to investigate various performance-analysis tools and recommend the one I thought was best suited for our department's needs. This short document explains my methodology and recommendations.

Methodology

I downloaded evaluation copies of the following three performance-analysis tools:

- Kaneval Eval v8.2 (made by Dexco Unlimited)

- Perf Burner v3.1 (made by Dexco Unlimited)

- PerfRomance v5.0 (made by Carambola Software)

I used all three tools to evaluate our latest software release, going through the documentation and trying out various features.

Recommendations

I strongly recommend PerfRomance v5.0 over the two Dexco Unlimited products for the following reasons:

- PerfRomance is the easiest product to learn.

- PerfRomance generates the clearest reports. (I've attached some comparable reports.)

- PerfRomance has the lowest cost.

In a parallel universe, Sam researched the same software tools. However, instead of writing up his recommendation as a formal document, he composed the following e-mail:

```
To: EngTeam
Subject: Recommendations
-------------------------------------------------------------------------------------

PerfRomance was very clean compared to the other two.
--Sam
```

Sam's e-mail message is, admittedly, concise, but the message is fuzzy. Furthermore, Sam's message doesn't contain any supporting detail. Engineers are critical thinkers who require substantiation of any claims.

Something strange happens to people when they compose e-mail. Even if an engineer brings great discipline to the art of writing formal documents, that same discipline usually vanishes in the writing of e-mail. When composing e-mail, people get very sloppy. Many busy people write e-mail as a stream-of-consciousness exercise, just blasting out messages. Some of my own e-mail messages have as much coherence (and I'm a professional writer) as the chaotic brain pattern I experience just before falling asleep.

People have much lower expectations from e-mail than from formal documents. Consequently, many people write e-mail to those low expectations, forfeiting all the knowledge they have about good writing. The first rule of writing work-related e-mail is, therefore, as follows:

> Use the same discipline in writing e-mail as you do when writing formal documents.

The preceding rule will sound trite until you give it a try. The same discipline includes running e-mail through a spell-checker, capitalizing normally, and even putting the pronoun *I* in uppercase.

Note that I'm only talking about work-related information. I'm not saying that messages like the following deserve any discipline at all:

```
To: EngTeam
Subject: Lunch
--------------------------------------------------------------------------------
Lunch at Mary Chung's at noon today?
--Sam
```

Before Hitting the Send Button...

If I were king of the world and the world had fallen into serious economic collapse, I would institute the following rule, which I believe would increase gross world product immensely:

> You may not send any e-mail until you have first reread it.

When rereading your e-mail, always ask yourself the following question:

> Will the recipient(s) understand this message?

For example, suppose Sam (based in London) sends out the following message to Nicole (based in San Francisco), who has just joined the project:

```
To: Nicole
Subject: status reports
---------------------------------------------------------------------------

status report needed by end of day tomorrow
```

At first glance, the preceding e-mail message seems innocuous enough. However, poor Nicole will probably not understand this message for the following reasons:

- **Like many e-mail messages, this message is written in passive voice.** In fact, the sole sentence is missing a pronoun altogether. Consequently, Nicole cannot be quite sure whether she needs to produce the status report or Sam needs her input on his own status report.

- **Since she just joined the project, Nicole probably has no idea what kind of status report is required.** Sam should send Nicole a status report template or an example status report.

- **The phrase *end of day* is ambiguous.** Sam's end of day (in London) comes quite a few hours prior to Nicole's.

- **Engineers and scientists have an extremely low tolerance for empty bureaucratic exercises.** Since Sam does not explain the purpose of the status reports, Nicole might not put the appropriate amount of effort into producing them.

- **The message is missing uppercase letters and punctuation.** I'm not a literary scholar, but I'm reasonably certain that most of the best writers[1] start sentences with an uppercase letter and end them with a period.

As is often the case with e-mail, Sam's original message omitted critical details and lacked any sort of empathy with the recipient. A brief moment of reflection on Sam's part would have solved the problem. Sam would do better to send the following e-mail instead:

```
To: Nicole
Subject: Please send me a status report

-----------------------------------------------------------------------------

Please write a status report and send it to me by 10:00 am (your time)
Thursday, Nov. 30. Your status report should contain the following two
sections:

   * "What I Accomplished This Month"
   * "What I Plan to Accomplish Next Month"

For each section, provide a bulleted list containing three or four items. I
like brevity, so try to keep each item down to a sentence or two. I've
attached a sample status report.


Everyone in my group writes a status report on the last working day of
every month. I use these status reports to produce a monthly status report
for my own manager. In addition, I store all your monthly status reports
and use items from them in your annual salary review.


If you have any questions, don't hesitate to send me an e-mail or call me.
  --Sam
```

---

1. One of the primary reviewers of this book (Chris Sawyer-Laucanno) published a biography of the poet e.e. cummings, who famously eschewed capitalization and sentence-ending punctuation in his poems. However, in e.e. cummings' personal letters (the e-mail of his time), he did use traditional capitalization and punctuation marks.

After the First Miscommunication...

When composing e-mail, people are often blunt. They type things in e-mail that they would never say in person. People are much wiser (or more cowardly) in face-to-face or telephone meetings than they are while composing e-mail. Therefore, before hitting the Send button, ask yourself the following question:

> Would you feel comfortable saying the text in this e-mail message in a face-to-face meeting with the recipients?

The preceding question is a reasonable starting point for e-mail sensitivity; however, you really have to take it a step further. After all, in face-to-face meetings, listeners attune to a whole range of social cues, including body language, facial expressions, and vocal inflections. In a face-to-face meeting, attendees know whether you are being sarcastic, but e-mail recipients might easily miss the sarcasm.

Emoticons

Emoticons are graphics that come in either of the following two forms:

- new-fangled graphic icons, such as the following which are quite popular in instant messaging:

- old-fashioned collections of punctuation marks, such as :)

The oft-stated purpose of emoticons is to provide emotional context for instant messaging and e-mail messages, since words alone allegedly cannot do the trick. For example, notice how the following emoticons defuse a potentially sticky situation:

```
You're an idiot. :) You have no idea what you're talking about. :)
Your experimental design is juvenile. :)
```

My initial response to emoticons was that writers have somehow managed to convey emotional context through words alone for several millennia. (Astonishingly, the works of Tolstoy, Shakespeare, and Cervantes do not contain a single smiley face.) I now realize that emoticons can be a valuable means of expression in e-mail and instant messaging *if* all recipients share a common understanding of their meaning. Note that many older readers are not well versed in emoticons and may find them annoying and confusing.

Do not place emoticons in any formal or legal written communication,

Miscommunication in e-mail often generates rage. For example, consider the following e-mail message, which does not successfully convey the intended idea:

```
To: EngTeam
From: Quincy (QA Manager)
Subject: Performance Testing on Project Sea Grape
--------------------------------------------------------------------------------
I ran performance tests on Project Sea Grape over the last week. I found
performance at the expected levels, except for the Batch Processing Module.
```

The preceding message angered Tim, who wrote the Batch Processing Module. He was not only angry, but he was embarrassed because Quincy had sent out the e-mail message to the entire engineering team. Tim responded in the following way:

```
To: Quincy
CC: EngTeams
From: Tim
Subject: RE: Performance Testing on Project Sea Grape
--------------------------------------------------------------------------------
There is nothing wrong with the performance of the Batch Processing Module.
In fact, it outperforms earlier releases by a factor of three. How did you
test it? Do you have any previous experience writing tests on Batch
Processing Modules?
```

In typical work environments, Quincy and Tim would get into a rapidly escalating e-mail war, with each side accusing the other of greater and greater ignorance. However, Quincy is not a typical engineer. Instead, he obeyed the following principle:

After the first miscommunication, stop sending e-mail. Use other media instead.

Quincy called Tim on the telephone to apologize. Quincy had meant that the Batch Processing Module performed far *better* than expected. Tim was relieved. Quincy then sent out a clarification to the entire EngTeam mailing list.

Scientists and engineers are more sensitive than stereotypes suggest.

Summary of E-Mail

When reviewing an e-mail message, just prior to clicking the Send button, ask yourself the following questions:

- Will the recipient(s) understand what you are trying to communicate?

- Does your e-mail message contain a relevant subject line? A good subject line not only makes sense initially but will continue to make sense in a few months. Note that many people now use their e-mail system as a sort of database management system for their work life.

- Have you edited the e-mail message?

- Are you sending this message to the appropriate recipients? Are you sending it to anyone who will be annoyed by receiving irrelevant e-mail? Are you excluding recipients who will feel left out if they learn that you didn't send them a message?

- Is the e-mail the right length? Recipients are often annoyed (and will ignore) lengthy e-mail. Conversely, make sure the e-mail is long enough to cover what it needs to.

- Could a recipient misunderstand your e-mail and get angry?

- What is your emotional state? If you are responding angrily to something, consider handling the situation in person or via the telephone instead of through e-mail.

- If you are sending e-mail to people in another culture, examine the message for the following:

  - Does your e-mail message contain slang that recipients in other cultures will not understand? In this case, jargon is okay, but slang is not. (See "Native Language" on page 13 and "Native Culture" on page 15.)

  - Could this e-mail message offend someone, even inadvertently, in another culture?

  - If English is a second or third language to the recipients, does your e-mail contain long, complex sentences? Your message should contain short, simple sentences.

  - Are you requiring that a recipient respond through e-mail? A recipient who is relatively unfamiliar with English might feel embarrassed to respond in written English to a large mailing list. Consider allowing the recipient to send an e-mail message directly to you, and then tactfully indicate that you will edit any grammatical or spelling problems before transmitting the edited message to the mailing list.

SECTION 4

Editing and Producing Documents

After completing the first draft of a document, you are not even close to being done with the job. This section explores how to edit properly and generate a professional-looking document.

CHAPTER 18

Editing and the Documentation Process

E ngineers and scientists generally spend more time reviewing other people's documents than writing their own. For this reason, it pays to learn a ~~little something~~ lot about editing and the documentation process.

Editing is the art of ~~humiliating writers~~ critiquing documentation to improve its accuracy and clarity. Generally speaking, editors falls into two categories:

- technical editors (often called technical reviewers)
- literary editors

Technical editors are fellow engineers and scientists searching for technical errors, logical fallacies, and ~~outright lies~~ mathematical mistakes. Good technical editors ~~live only to convict documents of perjury~~ save writers from publishing embarrassing documents. Good technical editors are ~~the bane of my existence~~ a writer's best friend.

Literary editors are experts in grammar, spelling, and technical writing who, typically, are neither scientists nor engineers. Literary editors fall into the following two categories:

- developmental editors
- copy editors

Developmental editors are writing experts who work at the macro level. They suggest ways to improve organization and writing style. For example, a developmental editor might suggest rearranging chapters or converting passive sentences into active voice.

Copy editors work at the micro level. They pore ~~obsessively~~ diligently through every letter of every blasted word looking for ~~nit-picky~~ spelling and grammatical errors. Copy editors are ~~stereotypically bespectacled women with hair pulled into a tight bun~~ highly attractive people who must be respected because we always get the last word, Rosenberg!

Editing: What Is It Really?

Most technical people believe completely in the veracity of the following formula:

> effective editing = finding mistakes

Unfortunately, the preceding formula is flawed. A more accurate formula is as follows:

> effective editing = getting writers to fix mistakes

The editor/writer relationship is inherently difficult. No one likes to be criticized. As an editor, you can find error after error, but your findings are worthless if any of the following are true:

- The writer believes that you don't know what you are talking about.

- The writer feels insulted by your comments.

- The writer has had dealings with previous editors and has trouble distinguishing you from them.

- The writer strongly believes that his or her way is better than yours.

For the preceding reasons, the foundations of successful editing are as follows:

- Building credibility so that the writer will take your comments seriously.

- Sharpening your diplomatic skills to reduce the writer's defensiveness.

In an ideal world, the writer sees that it is the writing, not the writer, being evaluated.

Forgive me—I just reread the previous sentence and realized that I've sprung a ridiculous cliche on you. In my experience, virtually no one can divorce one's writing from one's ego. Many professionals see their writing as an extension of themselves and get quite defensive when anyone criticizes it. Many professionals perceive editors as stumbling blocks—censors of their creative freedom. To many professionals, editors seem intent on removing "all the good parts." Good editors, meanwhile, are just doing their job, which is to uphold truth and improve documentation.

Ultimately, if things goes well, all artifice is removed. The writer sees the edits as valuable and respects the editor's improvements. Documents become more accurate and clearer. Somewhere, in a tiny, seldom-visited place in the writer's ego, the writer whispers "thank you" to the editor.

Technical Editing a Peer's Work

When performing a technical edit on a peer's work, follow the guidelines on this page.

Provide Positive Reinforcement

You should provide plenty of positive reinforcement in your comments. If a paragraph is well-written, say so. If a section is accurate and clear, say so. You'll be amazed at how effective even a few positive comments are. A spoonful of sugar really does make the medicine go down. Starting off your comments with a positive global comment can make all the difference.

Search for Omissions

Good technical editors search not only for technical mistakes but also for technical omissions. Technical editors often focus on what is present and neglect what is missing.

Avoid Vague Comments

Bad technical editors provide vague comments; good technical editors provide clear comments. For example, the following comment is worthless and rude:

> No!

The preceding comment only becomes useful when followed by an explanation of what is correct; for example:

> You were close—actually, the subtropical ridge is usually centered near latitude 32.

Suggest Replacement Text

To increase the odds of a writer accepting a suggestion, good technical editors suggest replacement or alternate text. For example, the following comment does not provide any replacement text, so a hurried writer might skip over it:

> This has to do with the subtropical ridge. Ask Michelle to explain.

The preceding suggestion necessitates some time-consuming research. The following comment has a far greater chance of being incorporated:

> I'd suggest something like the following: "Tropical cyclones typically curve around the perimeter of the subtropical ridge."

Technical Editing a Superior's Work

If you are editing the work of a superior (either a manager or a more senior scientist), follow the guidelines on this page.

Provide Thanks

Provide a quick preface to your edits, thanking your superior for the opportunity to edit, preferably without sounding too obsequious. For example, the following is just too gushy:

> It is a thrill to work with someone as famous as you.

However, a comment such as the following is appropriate:

> I learned a lot about artificial insulin research from reading this paper.

Make it clear that you would be honored to review subsequent drafts of the document.

Phrase Edits as Suggestions

Phrase your edits as suggestions whenever possible. Obviously, a misspelled word is a misspelled word and should be handled as a straightforward correction. However, you should handle more subjective matters with care. For example, the following is far too direct for a superior:

> Eliminate the opening sentence.

Rephrasing an imperative order as a question or suggestion will serve your career better; for example, consider the following improvement:

> How about if you omit the opening sentence and use that strong second sentence as your opening?

Copyediting a Colleague's Document

A copy editor is a wonderful resource that many organizations simply cannot afford. Sometimes, *you* must act as the de facto copy editor on a project because no one else is available. Sometimes you must take that critical last look at a document to catch any embarrassing mistakes before outsiders do. If you find yourself in this situation, here are a few suggestions for performing a quick copy edit:

- View the draft in the same medium as the target audience will. For example, if the target audience will read the document in PDF, review a PDF version of the draft, not a paper version.

- Check the table of contents. Does it include all the chapters and top-level section headers?

- Check all the headers and footers in the book. Are they consistent? Are page numbers displayed? Are any pages missing? If this is a book, do all chapters start on an odd-numbered page? Do the page numbers sequence correctly? (Don't laugh—page numbering mistakes are common, particularly in the transition from the table of contents to Chapter 1.)

- Get access to the word-processor sources for the document. Run a spell-checker on them. If a grammar-checker is available, run it. Most spell-checkers don't check for context. For example, if a writer used *there* instead of *their*, most spell-checkers won't report a mistake.

- Check tables and figures. Do they all contain captions? Do all titles and captions match up with their corresponding tables and figures? (In other words, are any captions in the wrong place?) Do all tables and figures have a numerical label, such as Figure 11-2 or Table 7-1?

- Verify that you placed trademarks or registered trademarks correctly. Consult with your corporate attorney, if necessary.

- If you only have enough time to review a tiny section of the document, evaluate the section or chapter that will draw the most readers.

Spreading the Wealth

Suppose a writer asks 10 technical editors to review a 300-page document. In all probability, only one or two of them will make it past the first 50 pages. Technical editors have limited time and limited attention spans. To improve coverage, the wise writer assigns different sections of the book to different technical editors.

Copyediting Your Own Document

Foxes should never guard the chicken coop, inventors should never act as their own quality assurance department, and writers should never edit their own work. Nevertheless, when no one else is available, you must become your own best editor. This is a suboptimal situation; you are simply too familiar with your own work to do a good editing job, so you will skip over mistakes that would have been obvious to another reader.

One of the best tricks for editing your own work is to read it aloud. This technique helps you find missing words and awkward sentences. It also does a great job of catching run-on sentences. (When you start to run low on oxygen, you've found a run-on.)

Reading aloud generally won't help you find spelling mistakes, but a spell-checker certainly will.

Another technique—one that sounds ludicrous but that some professional editors swear by—is to "read" the text backwards. Actually, you don't really read the text as much as you view it. The theory is that, when reading forwards, your mind coasts rapidly and carelessly through familiar text. By viewing the same passage backwards, you suddenly see it in a different way. Obviously, this technique isn't for everybody.

You will generally find more errors in passages that you wrote a while ago than in passages you recently wrote. An old saw—one that has some merit—suggests throwing your first draft in a drawer and forgetting about it for a while. If you are writing a lengthy document, consider putting yourself on a delayed editing schedule, where you edit a chapter six weeks after you've written it.

The preceding techniques are emergency stop-gap measures. Good writing, like good engineering, ultimately requires input and feedback from multiple parties.

Media for Technical Editing

Some engineering teams edit paper copies of documents. Other teams edit online.

For paper edits, editors take out their trusty red pen (nicknamed "Satan") and write comments directly on the hard copy. The writer eventually scoops up all the marked-up copies and incorporates the revisions. On the second round of editing, the writer should return the editors' original comments so that they can verify changes. The advantages of this approach are as follows:

- It is simple to implement.

- Many professional writers prefer this because it gives them more control.

- If the document being reviewed will be delivered to readers in hard copy, then reviewers see the document as readers will.

In recent years, many teams have switched to online editing. The technology for online editing ranges from fairly simple (each reviewer marks up a separate copy of the source document) to rather elaborate (using collaboration software, reviewers can see and comment on other reviewers' comments). The advantages of online editing are as follows:

- Most technical reviewers can type much faster than they can handwrite. Since the keyboard is mightier than the pen, reviewers tend to supply comments that are lengthier and more useful. Furthermore, writers never run into problems reading illegible handwriting.

- Depending on the technology, writers can sometimes simply indicate that they accept the change request, and it will be automatically implemented.

- Online editing saves paper.

Most high-end word processors provide a change-bar feature, which automatically flags any text that has changed since the previous version. Whether editing hard or soft copy, the change-bar feature helps focus editors' energy on the second pass.

Bug-Tracking Systems

Most engineering organizations use a bug-tracking system to monitor defects in a product. Many organizations also use these systems to monitor defects in documentation. The benefits of such a system are pretty much the same, whether you are tracking engineering or documentation defects. Namely, the key advantages are as follows:

- Unlike human memory, a bug-tracking system won't forget bugs.

- A bug-tracking system lets managers prioritize bugs to help focus effort on the most serious defects.

- A bug-tracking system records fixed bugs to prevent duplication of effort.

The disadvantage of a bug-tracking system for monitoring documentation defects is that it sometimes takes a disproportionately long time to open a new bug report. For example, a person who finds a misspelled word might avoid reporting it through a bug-tracking system if it takes five precious minutes to open a new bug report. After all, it is oh-so-much-easier to e-mail the writer about the mistake. I'm not saying that the easy way out is virtuous, but the goal is to detect and correct as many bugs as possible, and the bug-tracking system might hinder that goal.

I recommend using a bug-tracking system to report and monitor documentation bugs *after* the initial release of the document. Prior to the initial release, I recommend using the hardcopy or online editing techniques noted in this chunk.

QUANTUM LEAP
To reduce the number of documentation defects in early drafts (no matter how they are reported), always put your best effort out for review. In other words, spell-check and reread your document prior to submitting it. Too often, technical editors focus their reviewing energy on easy stuff (such as misspelled words) rather than on substantial technical issues. To keep technical editors focused, give them less low-hanging fruit to harvest.

A Process for Editing

Editing is essentially a quality assurance process for documentation. Therefore, the wise project team approaches documentation as an engineering process requiring a set of checks and balances. The process for creating documentation should ideally contain the following well-defined phases:

1. Writer creates documentation specifications.

2. Writer generates Draft 1.

3. Literary editor performs a literary edit on Draft 1.

4. Writer incorporates literary edit comments and generates Draft 2.

5. Team performs a technical edit on Draft 2.

6. Writer incorporates technical edit comments and generates Draft 3.

7. Team performs a technical edit on Draft 3.

8. Writer incorporates technical edit comments and generates Draft 4.

9. Copy editor performs a copy edit on Draft 4.

10. Writer incorporates copy edit comments and generates final document.

The preceding assumes a perfect world filled with plenty of time and resources. Unfortunately, the perfect engineering world is as attainable as a perfect vacuum. Many engineering teams skip plenty of the preceding steps to allow the writer more time for writing.

Enough Time to Review

Common sense suggests that the quality of edits and reviews increases as you give reviewers more time. However, common sense does not apply here. An old managerial adage states, "Work contracts or expands to fill the time allotted." My experience suggests that this adage is dead-on. For example, reviewers allotted three days to edit a 150-page document will do just as good a job as reviewers allotted three weeks. Naturally, there are limits—you can't expect someone to review a 500-page book in a day.

Technical reviewers often generate conflicting comments. When this happens, get all reviewers together over the phone or face-to-face. Don't let reviewers debate this sort of thing in e-mail as the discussion often turns contentious and unproductive. People's voices are far more polite than their e-mail messages.

Beta Tests for Documentation

In a beta test, real-world customers play with an almost-finished product and tell the inventors what they like and what they don't. Good engineers use this valuable input to improve their products.

Good engineering organizations also beta-test documentation. Unfortunately, most engineering organizations don't beta-test documentation optimally and end up with worthless comments such as, "The documentation was okay." If you want worthwhile beta-test comments on documentation, you have to go after those comments aggressively.

If the documentation is distributed in HTML format or as online help, build a documentation feedback button onto every page. When a beta-tester presses this button, a form should pop up asking for suggestions. Keep this form extremely simple; don't weigh it down with lots of extraneous information such as the beta-tester's title. In fact, the best form simply asks for suggestions, period.

If the total number of Best-testers is small, try to identify each individual who will read the beta documentation. At the beginning of the beta test, contact beta-testers and determine whether they are responsive to further questions about the documentation. (If you can get them to agree at the beginning, you'll be more likely to get good feedback later.) Assure beta-testers that all their feedback (good and bad) is quite helpful. Then, keep the lines of communication open. Get feedback while an idea is fresh in the mind of a beta-tester, so an instant medium such as IM is just about perfect.

Don't rely on surveys at the end of a beta-test to get documentation feedback. A closing survey will typically garner only general comments. For example, if a respondent rates your documentation as a 4 on a scale of 1 to 7, how exactly would you change your documentation? If your organization insists on a closing survey, try to get respondents to identify missing topics.

If your organization is financially blessed, offer financial carrots to beta-testers who provide significant feedback.

> ### Determining Where to Spend Documentation Resources
> When communicating with beta-testers, make sure you determine which documents they read most often. Then, after beta-testing finishes, focus your documentation resources on improving these documents. Fish where the fish are biting.

Summary of Editing and the Documentation Process

When performing any sort of edit, follow this general advice:

- Remember that you are editing the work of a fellow human and that most humans are sensitive to criticism.

- Find something to praise about the document, even if it's really bad; if there is actually something good about the document, well, all the better.

When performing a technical edit (technical review) on a document, focus on the following:

- Forget that you are you; instead, imagine that you are someone in the target audience for this document. Determine what *that* person would need to know.

- Perform any documented steps just as a user would. For example, if this is a software installation manual, find suitable hardware and use the documentation to install the software. Don't read between the lines—treat the text literally, doing exactly what it tells you to do.

- Determine what is missing from the manual.

- Provide clear criticism so that the writer knows how to fix the problem.

When performing a literary edit or a copy edit on a document, focus on the following:

- Follow the advice in Section 2 of this book, which details best technical-writing practices. Pass along this advice (diplomatically) to the writer. For example, identify passive-voice sentences that should be changed to active voice.

- Review captions to make sure that they are informative and accurate.

- Search for context problems that spell-checkers would miss.

- Identify passages that are unclear.

- Prior to editing, jot down a few topics that you expect to find in the book. Then, make sure that you can find all of them in the index and in the table of contents.

CHAPTER 19

Fonts and Typography

Although it is difficult to pick the best tourist attraction in New Jersey, my personal favorite is its never-ending supply of diners. For those of you who don't vacation in New Jersey, a diner is a friendly, inexpensive restaurant that features book-length menus, often with hundreds of entries. Sadly, the massive menus overwhelm many patrons who end up selecting something more dorky[1] than fulfilling[2].

Every modern computer system provides dozens, even hundreds, of fonts. Most computer users ignore this cornucopia and stick with the safe defaults chosen by their word processing software. This is a real pity. After all, fonts can give gravity to a serious document or leaven a casual document. Picking the right fonts can make your documents look more professional, just as picking poorly can detract from your message.

On the Other Hand, Mashed Potatoes Are Comforting

Fonts don't have to be exotic to be interesting. I don't want to suggest that you should set a serious technical document in a novelty font such as Papyrus or Park Avenue just because it would give your document a distinct look. Novelty fonts can punch up a birthday card or an advertisement, but they have no place in the kinds of scientific documents that you are writing.

Until a few decades ago, a chapter on fonts would not have appeared in a book for mere mortals. In the old days, engineers and scientists wrote the words and left all font decisions to trained printing professionals. Nowadays, though, anyone can conjure font changes with a few mouse clicks. This chapter helps you click wisely.

---

1. Burgers and fries.
2. Latkes with apple sauce. Try the cheese blintzes, too.

Serif and Sans-Serif Fonts

If you look very closely at the words in this sentence, you'll see tiny horizontal and vertical notches popping out from the ends of most of the letters. Just to make it a little more obvious, I've blown up a lowercase *p* in Figure 19-1.

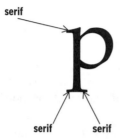

FIGURE 19-1   This lowercase *p* contains three serifs.

Notice the tiny horizontal lines at the base of the *p* and the tiny little hook in the upper left. These tiny lines are called **serifs**. Some fonts have them, and some don't. Fonts in which most of the characters have serifs are called (not surprisingly) **serif fonts**. Fonts in which none of the characters have serifs, such as the character shown in Figure 19-2, are called **sans-serif fonts**.

FIGURE 19-2   This lowercase *p* contains no serifs.

Table 19-1 summarizes three popular serif fonts. The *Weight* column refers to the relative thickness of the lines and curves constituting each character. For example, printing the letter *o* in a heavy-weight font expends more ink than printing it in a light-weight font. Many of these fonts provide a boldface version that is exceptionally heavy.

TABLE 19-1 A Few Popular Serif Fonts

Name	Example	Weight	Comment
Garamond	The quick brown fox	Light	This font suggests elegance and precision. It features ornate serifs (look closely at the *T*), which strike some readers as old-fashioned.
Palatino-Linotype	The quick brown fox	Medium	This is a good choice for technical documents. Characters in this font are a bit wider than in the other two fonts, so text consumes more space.
Times New Roman	The quick brown fox.	Heavy	This font also makes a good choice for technical documents, but because of its ubiquity, documents in Times New Roman do not stand out.

Table 19-2 summarizes three popular sans-serif fonts.

TABLE 19-2 A Few Popular Sans-Serif Fonts

Name	Example	Weight	Comment
Arial	The quick brown fox	Medium	Microsoft developed this font for online use, so it looks excellent online; however, it is not great for hard copy.
Benton-Gothic	The quick brown fox	Light	This font looks very sharp in hard copy but a little smudgy online.
Verdana	The quick brown fox	Medium	This font gives hard-copy documents a distinctive look. Notice how wide it is relative to the other two fonts in this table. If space is at a premium, this font would not be a good choice.

Many systems use Arial and Times New Roman as the default fonts, so users with typographic inertia often just go with these fonts. Both fonts are sturdy choices. However, since they are ubiquitous, documents that use them do not stand out. If you want to imbue a document with a distinctive look or brand, then you must choose nondefault fonts.

Fixed-Width versus Variable-Width Fonts

All the fonts examined so far—whether serif or sans serif—have been **variable-width fonts**. In a variable-width font, each character occupies only the horizontal space it needs. In the following example, notice how B and i snuggle up tightly:

> BiBi

Every character in a **fixed-width font** (also called a **monospace** font), whether brawny or svelte, consumes exactly the same amount of horizontal space. Thus, naturally skinny characters end up with a lot of white space around them. For example, in the following example, notice how much blank space is between each pair of letters:

> `BiBi`

Fixed-width fonts use horizontal space less efficiently than variable-width fonts. Table 19-3 describes two popular fixed-width fonts.

TABLE 19-3 Two Popular Fixed-Width Fonts

Font Name	Example	Comments
Courier New	`The quick brown fox`	This is the classic, universally available fixed-width font. Similar fonts include Courier.
Lucida Console	`The quick brown fox.`	Notice that Lucida Console is wider than Courier New. Many readers find Lucida Console more legible than Courier New; however, Lucida Console is not as universally available as Courier New.

Fixed-width fonts are generally less readable than variable-width fonts. In addition, large blocks of text in fixed-width fonts don't look good. Nevertheless, you should use them to display the following:

- programming source code or command lines

- mathematical or scientific equations (although you probably need a special font for many equations)

- values that must line up, such as in a spreadsheet or database

Serif and Sans-Serif in Hard Copy

When producing hard-copy documentation, a simple formula for typographic success is as follows:

- Use a serif font for all body text, which includes the following:
  - paragraphs
  - lists
  - table cells
- Use a sans-serif font for all header text, which includes the following:
  - chapter titles and section headers
  - table headers
  - table and figure captions
  - figure callouts
  - page headers and footers, plus page numbers
- Use a fixed-width font for all special elements, which include the following:
  - equations
  - software commands or code

In other words, a document should typically contain a grand total of only three fonts—a serif font, a sans-serif font, and a fixed-width font. For example, you might use Palatino for all body text, Verdana for all header text, and Lucida Console for all special elements. If you were to use only a single font throughout the document, section headers and table captions would tend to hide. Following the three-font formula makes it easier for readers to find what they are seeking.

Serifs Are Sweet Noise on Paper

Many centuries ago, typographers discovered that serif fonts were more readable than sans-serif fonts. This is rather astonishing because serifs are essentially noise—extraneous particles strewn over a clean surface. (Can you think of any other instance in which adding noise to a system makes it more comprehensible?) Although sans-serif fonts look cleaner than serif fonts, the readability evidence in favor of serif fonts is overwhelming. Nearly every hard-copy novel is set in a serif font.

Serif and Sans-Serif in Soft Copy

When producing soft-copy documents, you should rely entirely on sans-serif fonts.

Serifs Strike Sour Chords Online

In hard-copy documents, serif fonts are more readable. However, in soft-copy documents, sans-serif fonts are more readable. The reason has to do with the resolution of the different media. Most hard-copy published documents are printed at 1,200 dots per inch. However, even the best monitors can render soft-copy documents at only 125 dots per inch or so. At such a low resolution, serif fonts look smudged, but sans-serif fonts still look clean and sharp.

Most people choose one of the following two strategies for using sans-serif fonts in soft-copy documents:

- Apply a single sans-serif font throughout the document.
- Apply two different sans-serif fonts; for example, use Arial for body text and Verdana for headers.

Even if you choose the latter strategy, you must still rely on additional characteristics (such as font size, indentation, and color) to differentiate body text from header text. Different sans-serif fonts generally look rather similar; most readers cannot easily differentiate Arial from Verdana. Using a different color is much more obvious than using a different sans-serif font.

Arial is a very popular choice right now for online documentation, particularly on machines running Microsoft Windows. That is because Arial was designed specifically for soft-copy documents. (Most fonts were designed for hard-copy documents.)

Mixed Media

Sometimes you must provide the same document in both hard copy and soft copy. In such cases, you should determine which medium is more popular and apply fonts accordingly. When hard copy and soft copy are equally popular, I usually use the rules for hard-copy documents. One other possibility is to generate two different versions from the same source. The only disadvantage to doing so is that page numbering might differ between the two versions, which can cause confusion.

Font Height

Font height is measured in a peculiar unit called a **point**, where

```
72 points = 1 inch
```

Given the preceding definition, you might expect a 36-point character to be exactly 0.5 inches tall. You silly logician, you! In fact, every 36-point character is far shorter than 0.5 inches. However, measuring from the bottom of the font's lowest letters (for example, *g* or *j*) to the top of the font's tallest letters (for example, *Z* or *I*) does yield a height of exactly 0.5 inch. See Figure 19-3.

FIGURE 19-3   A 36-point font is 0.5 inches from the lowest point to the highest.

Best Font Sizes for Hard Copy

The following are a few general guidelines for setting font sizes in hard-copy documents:

- **Set all body components (paragraphs, lists, and table cells) to the same font size.** The size you pick should be somewhere between 10 and 11 points, inclusive. For example, don't set paragraphs to 10 points and bulleted lists to 11 points.

- **Sans-serif fonts look a little bigger than serif fonts of the same point size.** For this reason, set the following components one point smaller than body components:

  - table headers

  - table and figure captions

  - figure callouts

  For example, if you set your body components to an 11-point font, set your table headers to a 10-point font.

- **Leave big deltas between different section header levels.** I like to leave a difference of at least three points (four points is better) between the font sizes of a first-level header and a second-level header. For example, readers can easily distinguish between an 18-point first-level header and a 14-point second-level header, but the difference between an 18-point and a 16-point font is much harder to distinguish.

Larger Fonts Appeal to Older Readers

One of the central theorems in technical communication is to be clear. For this reason, good technical communicators consider the age of their readers when making font decisions. When writing for the over-40 set, never pick a body font smaller than 10 points. Older readers greatly prefer an 11-point body. Remember— your readers can't click a menu to make text bigger in a hard-copy book.

Best Font Sizes for Soft Copy

PDF and HTML are the two most popular formats for online documentation distribution.

When creating PDF documents, *absolute* font sizes are not important since readers can easily change the effective font sizes through a simple menu selection. Nevertheless, *relative* fonts sizes are still important, so the size guidelines for hard copy are relevant for PDF documents.

When creating HTML documents, you must choose between one of the following strategies for setting font sizes:

- Set font sizes explicitly through HTML tags or through cascading style sheets (CSS).

- Do not set font sizes; instead, let users set them through browser controls.

The first strategy—setting font sizes (and other font characteristics) explicitly—offers authors the greatest control. For example, with a CSS, you can tell the browser to render all paragraphs in 11-point Arial. However, this control is somewhat illusory because you cannot control the screen resolution at which readers will view the document. For example, 11-point Arial looks sharp on an 800 × 600 screen but becomes intolerably tiny on a 1600 × 1200 screen.

The second strategy—letting users pick font sizes—is generally preferable for technical documents. This strategy permits users the luxury of adjusting the font to meet their eyes' needs. In addition, this strategy is much easier to implement than a CSS.

Italics and Boldface

Most fonts provide *italic* and **boldface** (or just **bold**) variants. Both draw attention to a particular word, so when would you italicize and when would you bold? This page provides a few guidelines. Note that these are only guidelines, not hard-and-fast rules. Whatever guidelines you choose, apply them consistently within a document or documentation set. For example, if you introduce new terms with boldface in one chapter, don't introduce new terms with italics in another chapter.

Boldface

Boldface stands out more than italics. Therefore, if you really want to draw readers' attention to a word or phrase, put it in boldface. Don't overuse bold—a little bold goes a long way. If a page contains too much boldface, the impact of boldface fades. Note that many headers appear to be in boldface even when they aren't. (Large fonts often look bold.) I recommend using boldface for the following passages:

- when introducing new terms in a paragraph, as in the following passage:

Polyprotic acids contain multiple acid hydrogen in each molecule.

- in the portion of a command-line dialog that a user enters verbatim; for example, notice that the user literally enters pr1me and 23 in the following passage:

```
$ prime
Enter an integer: 23
23 is prime.
```

Italics

I recommend using italics for the following:

- the title of a book, article, or other copyrighted source

- a foreign word or phrase not found in an English dictionary

- in software documentation, any nonverbatim word or passage that the user must enter; for example, in the following command, the user must enter an integer (not the literal word *integer*):

```
dexprime integer
```

Note that italics show up badly online, so many Web writers use boldface for emphasis rather than italics.

Consistency and Convention

Whatever fonts you choose, apply them consistently. For example, if you start off with Palatino for all body text, then stick with it throughout the entire document. In addition, you should also apply boldface, italics, and fixed-width fonts consistently.

How do you make sure that you apply fonts consistently? The key is to work with a strict word-processing template[3] rather than blindly pressing the `Italic` or `Boldface` button. For example, a good word-processing template should define components with names such as the following:

- `Emphasize`
- `NewTerm`

When a writer needs to emphasize a passage, instead of trying to remember whether to bold or italicize, the writer simply "tags" the text with the `Emphasize` component. In this way, all emphasized passages end up in the same font.

One of the benefits of the preceding approach is that it simplifies global changes. For example, suppose the `Emphasize` component is defined as italic. If a new editor arrives and decides that the corporate standard for emphasized passages should be boldface, it is a simple matter to implement this change just by redefining the `Emphasize` component.

Font Conventions

Many manuals contain a "Font Conventions" section identifying the purpose or usage of each font in the manual. For example, such a section might indicate that 10-point Courier bold represents text that the user must enter verbatim.

On one hand, "Font Conventions" sections remove potential ambiguities. On the other hand, if a manual requires such a section, then the fonts in that manual were probably not clear enough.

---

3. This is often called an **SGML** approach. SGML, or Standard General Markup Language, is a superset of HTML.

True-Type versus PostScript Fonts

Two different font technologies dominate the personal computer world. Almost all the fonts on a Windows PC or Macintosh rely on one of the following standards:

- True-Type fonts, which have the .TTF file extension on Windows-based PCs

- PostScript (also called PostScript Type 1 or just Type 1) fonts, which have the .PFM and .PFB extensions on Windows-based PCs

Which category is better? True-Type fonts get more use because Windows PCs come fully stocked with them. However, most professional graphic artists prefer the superior quality of PostScript fonts. PostScript fonts come in tandem files, with one file (the .PFM file) holding the screen version of the font and the other (the .PFB file) holding the printer version. In some cases, font creators produce both a True-Type and a PostScript version of the same font.

A single document should always use a single font technology. In other words, never mix True-Type and PostScript fonts in a single document. Mixing font technologies often causes printing problems.

Summary of Fonts and Typography

Ask yourself the following questions about the font choices in your document:

- What is the output medium for this document?
  - If the output medium is hard copy, have you used serif fonts for body and sans-serif fonts for headers?
  - If the output medium is a computer screen, have you used sans-serif fonts for everything?
- Are the fonts big enough to be legible?
  - Will older readers have trouble reading these fonts?
  - If this is an HTML document, will readers on high-resolution monitors be able to see these fonts? Can readers change the size of fonts?
  - If this is a hard-copy document, is the body text set in 10- or 11-point type? Are the headers noticeably larger than the body text?
- Are fonts used consistently throughout the document (and document set)?
  - Do the settings for first-, second-, and third-level headers change?
  - Do you use italics and boldface consistently?
  - Do you use consistent fonts for headers and footers?
- Are the fonts obtrusive? Fonts, like journalists, should tell the story without calling attention to themselves.

CHAPTER 20

Punctuation

Please forgive this grammatical intrusion, but many people have trouble using the following punctuation marks correctly:

- commas

- dashes and hyphens

- semicolons

- periods

- colons

- quotation marks

This appendix does not provide a comprehensive look at these punctuation marks; it only summarizes their most common uses within technical prose. If you want to learn every nuance, see a style manual, such as the *Chicago Manual of Style*.

Musical Punctuation

If words are notes, then punctuation marks are rests. Musically speaking, you might think of punctuation in the following way:

- A comma is a quarter-note rest.

- A dash is a half-note rest.

- A semicolon is a three-quarter note rest.

- A period is a full-note rest.

If sheet music isn't your thing, don't worry—all will become clearer in the next few pages.

Commas

English is a rather flexible language, and there is very little to stop you from placing a comma just about anywhere. A comma signals a short pause or division. If you are uncertain whether to use a comma, just read the sentence out loud and place a comma anywhere your voice rests briefly. For example, when reading the following sentence out loud, your voice should pause briefly just after the word *life*:

> Although carbon is one of the critical elements of life, a tiny percentage of carbon atoms are naturally radioactive.

Technical prose often contains conditional sentences. In a conditional sentence, use a comma to separate the condition from the consequence. For example, the following conditional sentence contains an explicit *if* and *then*:

> **If** the pressure drops by more than 10 mb in an hour, **then** a severe storm is on the way.

The previous sentence contains an explicit *then*, but in many conditional sentences, the word *then* is implied. In such sentences, the comma is critical to helping the reader identify the implied *then*. For example, in the following example, notice the comma after *hour*:

> **If** the pressure drops by more than 10 mb in an hour, a severe storm is on the way.

As noted in Chapter 7, technical communicators prefer bulleted lists to embedded lists. Nevertheless, if you do create an embedded list with more than two items, place commas after each element in the list except the final element. For example, in the following example, note the comma after the first element (*oranges*) and the second element (*lemons*):

> Three types of citrus fruits are oranges, lemons, and limes.

Good technical prose is built for speed; commas act as speed bumps. As a rule of thumb, sentences within technical prose should rarely contain more than two commas. Sentences containing more than two commas are ripe for editing.

Dashes and Hyphens

Dashes come in the following two flavors:

- em dashes (—), which are as wide as the letter "M" in a given font

- en dashes (–), which are as wide as the letter "N" in a given font

The en dash is the grammatical fifth Beatle—most people just want it to go away. Its only common use in technical prose is in numerical spans, such as the following:

> The hybrid engine increases gas mileage 30–50%.

Like commas, em dashes are signals for the reader to pause. Readers pause longer for em dashes than for commas. Em-dashes often travel in pairs. The text between a pair of em dashes often provides a quick definition, as in the following example:

> Iodine is a halogen—a nonmetal with a single electron to donate—and a common allergen.

Hyphens

A **hyphen** is a very short dash (-), even narrower than the en-dash. Word processors automatically supply hyphens to split multisyllabic words at the end of a physical line. You explicitly supply hyphens to string together a group of unsplittable words. For example, the hyphens in the following example unite four words into one compound noun:

> Crank the handle as you would a **jack-in-the-box**.

The hyphens in the preceding example are essentially the opposite of commas; commas cause readers to slow down, but hyphens cause readers to speed up. In the preceding example, the reader mentally runs the hyphenated *jack-in-the-box* together as one word. You also use hyphens to join a multiword adjective preceding a noun; for example, consider the hyphens in the following example:

> The corpus callosum helps convert **short-term** memories into **long-term** memories.

Be careful—do not use a hyphen unless the modified noun follows the multiword adjective. For example, the following example correctly contains no hyphens because the noun *memories* does not follow either of the multiword adjectives:

> The corpus callosum helps convert memories from **short term** to **long term**.

Semicolons

Use a semicolon to connect two closely related, complete sentences. A semicolon and a period are essentially interchangeable; however, the semicolon suggests the marriage of a perfectly matched couple, almost as if the writer couldn't bear to separate the two thoughts with something as harsh as a period. In the following sentence, notice that the two sentences are so closely related that the writer could have reversed their order:

> Hydrogen is the only element without a neutron; each hydrogen atom contains only a proton and an electron.

Good writers often place transition words such as *however* and *therefore* immediately after a semicolon. Remember to place a comma immediately after the transition word. For example, consider the following:

> A stable helium nucleus contains two protons and two neutrons; **however,** several unstable helium isotopes contain more than two neutrons.

You may place a transition word in the middle of a sentence. In such instances, put a comma before and after the transition word, as in the following example:

> A stable helium nucleus contains two protons and two neutrons; several unstable helium isotopes**, however,** contain more than two neutrons.

You may place multiword transition phrases (such as *on the other hand* or *for example*) immediately following a semicolon, as in the following example:

> On one hand, Java enables truly portable programs; **on the other hand,** Java programs often run slowly.

Never place a conjunction (such as *and* or *but*) after a semicolon; for example, the following grammatically improper sentence requires a comma instead of a semicolon:

> Assembly language programs run quickly; **but** they are difficult to code.

Occasional use of semicolons helps your writing look more professional. Overuse of semicolons tends to look a little sophomoric.

Periods

As you know, you place a period at the end of most sentences. However, the humble period also serves a few other uses in technical prose, which this page describes.

Parentheses provide a splendid opportunity to misuse periods. If you place a full sentence within a pair of parentheses, then place the period just *inside* the closing parenthesis, as in the following example:

> Hurricanes have sustained winds of at least 65 knots. (In the Pacific Ocean, hurricanes are called typhoons.)

If you insert a phrase (not a complete sentence) within a pair of parentheses, then do not place a period within the parentheses. For example, notice that the period appears *after* the closing parenthesis in the following example:

> Hurricanes have sustained winds of at least 65 knots (74 miles per hour).

You place a period after an abbreviation but not after an acronym, as in the following examples:

> Dr. (abbreviation—ends with a period)
>
> DBMS (acronym—does not end with a period)

Exclamation Points

In recent decades, editors and pundits have vilified the exclamation point so much that writers who dare use it are now generally sneered at. The feeling among the literati is that exclamation points are a cheap trick to infuse excitement. Real writers, they say, build excitement through words. While I don't believe that inserting exclamation points is a capital crime (hey, I've put several exclamation points in this book), I'd recommend using them sparingly.

Colons

Within technical prose, use a colon to introduce a list, table, or figure that will immediately follow. For example, the following passage correctly uses a colon to introduce a bulleted list:

> Water has the following properties:
>
> - It is an excellent solvent.
>
> - It becomes less dense below 4°C.

If the list, table, or figure does not immediately follow its introduction, then terminate the introductory sentence with a period rather than a colon. For example, in the following passage, notice that the introduction occurs a few sentences prior to the start of the list:

> Water has the following properties. Note that water is the only compound to have these critical properties. Without the second property, most aquatic life would die.
>
> - It is an excellent solvent.
>
> - It becomes less dense below 4°C.

The sentence introducing a list should appear on the same page as the start of the list. However, it is okay for a table or figure to appear on a different page than the sentence that introduces it. In this case, the sentence introducing the table or figure should end with a period rather than a colon, even if there is no additional text after the introductory sentence. For example, compare the right and wrong ways of introducing a figure that appears on a different page:

> Figure 8 on page 24 illustrates packet passing in a token ring network. (right)
>
> Figure 8 on page 24 illustrates packet passing in a token ring network: (wrong)

Quotation Marks

In fiction, quotation marks bound the beginning and end of a spoken line, as in the following example:

> Woody said, "And, I think what we've got here is a dead shark."

Technical prose offers very few opportunities for such juicy dialog. However, quotation marks allow technical authors to bound an exact quote from another written source, as in the following example:

> Newton's First Law of Motion states, "Every object in a state of uniform motion tends to remain in that state of motion unless an external force is applied to it."

A solid alternative to quotation marks is to place verbatim text on separate lines, indented as follows:

> Newton's First Law of Motion states the following:
>
> > Every object in a state of uniform motion tends to remain in
> > that state of motion unless an external force is applied to it.

Some sources advise writers to indent only those quotes that are longer than ten physical lines. Personally, I also find it useful to indent much shorter quotes.

You may place quotation marks around a particular word or phrase to indicate that you are using that word in a special or nonstandard way. For example, the following quotation marks are a little wink to the reader to indicate that you realize that robots are gender-neutral:

> The robot was dressed in a tuxedo when "he" answered the door.

Avoid using quotation marks that might hamper clarity. For example, consider the quotation marks in the following sentence:

> To remove the locked file, issue the command, "rm *.lck."

The preceding example might fool the reader into thinking that the closing period is a required part of the command. When there is even a remote chance of confusing readers, remove the quotation marks and do one of the following instead:

- Mark the special text with a different font.

- Place the special text on a separate line.

The latter option is the better approach. For example, the following version removes any possible confusion about what command to enter:

> To remove the locked file, issue the following command:
>
> rm *.lck

Straight Quotes versus Curly Quotes

Many word processors provide the following two kinds of quotation marks:

- "straight quotes," which look a little like the number 11

- "curly quotes," which look a bit like the numbers 66 and 99

Generally speaking, you should use curly quotes. The only common exception is when documenting sample program code. For example, note that the following code contains straight quotes:

```
printf("The answer is %d.", my_answer);
```

 GLOSSARY

abstract   A brief summary of a lengthy proposal or lab report.

active voice   A sentence in which the subject acts on the object. For example, *Cheetahs chase zebras* is an active-voice sentence because the subject (cheetahs) verbs the object (zebras). Active- voice sentences are usually shorter, clearer, and more powerful than passive-voice sentences. In addition, most readers for whom English is a second language prefer active voice sentences. Compare to **passive voice**.

audience   The people who will read your document or view your Web site. The audience members for a Web site are also called *visitors*.

block diagram   A figure typically consisting of a set of rectangles with embedded labels. In some block diagrams, an artist draws arrows between the rectangles to connect related structures visually. In other block diagrams, the artist stacks rectangles vertically to symbolize a hierarchy.

bullet   The typographic mark—usually a filled circle or square—that denotes the start of a new element in a bulleted list.

bulleted list   A list in which each element begins with a bullet. Unlike a numbered list, the elements in a bulleted list have no specific order. In other words, if you rearrange the elements of a bulleted list, the list would still have the same meaning. Compare to **numbered list.**

business plan   A proposal aimed at getting money to start a new business or expand an existing business.

business proposal   A proposal aimed at selling a new idea (typically, for a new product) within an existing company.

callout   A combination of a textual label and a pointer (a line segment or arrow). A callout points to a particular part and gives it a name. Compare to **embedded label.**

caption   A brief description of a figure, typically prefaced by a number. By custom, captions appear just below figures. The captions associated with a table are sometimes called **titles.**

chunk   A short, discrete unit or lesson on a particular topic. Typically, each chunk starts at the top of a new page.

color blindness   A condition in which people (usually men of European background) have trouble distinguishing among certain colors.

conjunction   A word (such as *and* or *but*) that connects two related thoughts within the same sentence. Compare to **transition.**

context-sensitive help   A kind of online help system that automatically provides the relevant help information based on the user's current situation. With context-sensitive help, users do not have to search for the appropriate help file.

cookbook-style manual   A manual similar to a culinary cookbook in which each topic consists of a list of materials followed by a numbered list of instructions.

copy editor   A type of literary editor who looks for errors in spelling, punctuation, and grammar. Compare to **developmental editor.**

cover letter   A formal business letter that introduces an attached document (such as a resume or proposal).

CSS   Abbreviation for *Cascading Style Sheets*. A CSS enables Web developers to maintain a distinct and consistent look across an entire Web site.

design spec   In some industries, another name for a low-level technical spec.

developmental editor   A type of literary editor who acts as a sort of writing tutor to improve organization and writing style. Compare to **copy editor.**

doc project plan   A planning document that summarizes an entire documentation set to be written. A doc project plan explains how each manual will fit into a cohesive whole.

doc spec   A planning document that summarizes a single manual to be written. Doc specs usually define the manual's audience, length, and approach. Good doc specs also provide a detailed outline and schedule.

dynamic content   Program-generated Web information. Unlike static content, dynamic content can change depending on who is visiting the site or what a given visitor wants to do with the site.

elevator speech   A very short oral presentation to a venture capitalist or banker in which an entrepreneur describes an idea for a new company.

element   A numbered or bulleted item within a list.

embedded label   A label placed directly on the part of a figure that it describes. Compare to **callout**.

endnote   Scrap of information (often a citation) that appears in a note at the end of a document. Endnotes and footnotes typically contain the same sort of information, but footnotes appear on the source page, and endnotes appear at the end of the document.

establishing shot   A graphic that shows an entire scene or the exterior of an object. Photographers, cinematographers, and graphic artists like to present an establishing shot prior to graphics that provide details.

FAQ   Abbreviation for *frequently asked questions*. FAQs are a staple of newsgroups and other Internet communications. As the name suggests, a FAQ is a list of common questions and their answers.

fixed-width font   A font in which each character consumes exactly the same amount of horizontal space. Compare to **variable-width font**.

font   A graphically consistent set of characters.

footnote[1]

functional spec   In many industries, another name for a high-level technical spec.

---

1. A numbered note (like this one) that appears at the bottom of a page. Formal scientific and technical writing reserves footnotes for certain kinds of citations. However, more casual scientific and technical writing may also use footnotes for digressions. In general, technical documentation shouldn't contain too many footnotes.

guide   A manual more advanced than a tutorial, intended for more experienced readers. Good guides generally teach topics by presenting a series of examples, moving from moderately easy to more advanced ones.

header   A title within a document. The definition of *header* has shifted somewhat in the last couple of decades. Originally, a header was any title, including a chapter title. Now though, many people do not think of a chapter title as a header and believe that only those sections or subsections within a chapter can be called headers. Headers are hierarchic; thus, documents typically contain first-level headers, second-level headers, and so on.

Hello World example   The simplest possible example. A Hello World example should typically be the first example in any technical book. The phrase comes from the first example in a book called *The C Programming Language* by Brian Kernighan and Dennis Ritchie.

high-level technical spec   A detailed internal document aimed at a multiple departments (for example, sales, marketing, and documentation). A high-level technical spec summarizes a product about to be developed.

home page   The Web page intended to serve as a starting point for visitors to a Web site. All Web sites should contain a home page, although it is not a given that visitors will browse to the home page before viewing other Web pages. Compare to **secondary pages**.

HTML   Acronym for *Hypertext Markup Language.* The bread-and-butter encoding language of the Internet, consisting of simple tags (such as <LI> for a list) and plain text. HTML is a subset of a larger encoding scheme called SGML (Standard Generalized Markup Language).

imperative verb   A verb that acts as a command. For example, in the sentence *Put the cap on the tube*, the word *put* is an imperative verb.

jargon   The words, phrases, and acronyms that practitioners use when communicating among themselves but that are unknown to most nonpractitioners.

layout   The geometric arrangement of graphic elements (graphics and text) on a virtual or physical page. This is sometimes called *composition*.

literary editor   A person who reviews documentation with an eye toward the writing itself rather than the technical content. *Copy editors* and *developmental editors* are two types of literary editors. Compare to **technical editor**.

low-level technical spec A detailed internal document aimed at the engineers on a project who will be implementing a new product or technology. Such specs provide the blueprints for how the engineering team should implement the product or technology.

manual A document that teaches readers how to perform a task, use a product, or master a technology.

monospaced font Another name for fixed-width font.

nonverbal manual A manual that contains only graphics and no words.

numbered list A list in which each element is preceded by an integer. The elements must be listed in a particular order. If you can change the order of the elements in the list, then it should be a bulleted rather than a numbered list.

online help Generally, any documentation presented in soft-copy format. However, most people now reserve the term *online help* to mean a particular kind of soft-copy documentation that provides quick, concise explanations of how to accomplish specific tasks.

pace The rate at which a document presents information. A fast-paced document presents a lot of new information per page, while a slow-paced document presents only a little new information per page. Slow-paced documents spend more time explaining assertions than fast-paced documents.

page template Instructions that produce a consistent layout across a set of Web pages. Strict adherence to a good page template helps produce professional-looking Web sites.

parallelism Grammatical and logical consistency among all the elements in a list. Elements within a list should be completely parallel. For example, if the first element in a list is a singular noun, then all subsequent elements in the list should be singular nouns.

parenthetical clause A digression, example, or elaboration *within* a sentence. A parenthetical clause is bounded by a pair of commas, parentheses, or dashes.

passive voice A sentence in which the subject is acted upon by the object. For example, the sentence *Zebras are chased by cheetahs* is in passive voice because the subject (cheetahs) is verbed upon by the object (zebras). Use passive-voice sentences sparingly in technical and scientific writing. Compare to **active voice**.

PDF Abbreviation for *PostScript Display Format*. With the right software, authors can convert documents written in any word processor to PDF. Users with the right software (such as Adobe Acrobat Reader) can view or print PDF documents.

point In typography, a unit for measuring the height of fonts, as measured from the top of the highest letter to the bottom of the lowest. One point is 1/72 of an inch high. The point is one of the slipperiest units of measurement in the entire mathematical world; for example, every letter in a 12-point font is shorter than the expected 1/6 inch.

pointer In layout, anything that directs a reader's eyes. For example, an arrow is a pointer because readers will look in the direction that the arrow is pointing.

PowerPoint Software manufactured by Microsoft (and part of the Microsoft Office Suite) that helps organize presentations.

preproposal A brief summary—either written or oral—communicated to reviewers prior to creating a full proposal. Use the preproposal to test the waters.

proposal A written request to do something—for example, to perform research, start a company, or develop a program—in exchange for money or other resources.

Question-and-Answer (Q-and-A) Format A style of documentation in which each header is a question, which the body text answers.

readability quotient A mathematical formula for estimating the appropriate educational level of a document's audience. For example, a certain readability quotient might determine that a given document is appropriate for sixth graders.

reference manual A manual in which details are presented as a series of discrete topics. Reference manuals usually provide rather terse descriptions compared to guides or tutorials.

release notes A document popular in software products that typically describes the new features of a software release, plus any known bugs and bugs fixed in the current release.

research proposal A written request to research a problem in exchange for money or other resources.

rule A line that divides cells within a table or forms a border on the outside of a table.

run-on sentence   A sentence that is longer than it should be. Alternatively, a sentence that the writer should divide into two or three separate sentences.

sans-serif font   A font in which none of the characters are rendered with tiny little notches (serifs) at the endpoints. Sans-serif fonts are more readable than serif fonts in online documents.

screenshot   A digital picture of an image that appears on a computer monitor. Software documentation for end users, installers, and system administrators often features screenshots.

secondary page   Any Web page in a Web site other than the home page.

serif font   A font in which all or most of the characters are rendered with tiny little notches (serifs) at the endpoints. In hard-copy documents, serif fonts are more readable than sans-serif fonts.

SGML   Acronym for *Standard Generalized Markup Language*. HTML is a subset of SGML. Both are tagging languages, but SGML parsers are much stricter than browsers.

shading   In tables, a gray or colored background on individual cells or on rows. Many tables use alternate-row shading in which every other row is shaded and the alternate row is unshaded.

sidebar   A block of text and/or graphics displayed in a shaded background that contains an interesting digression off the main topic.

significance statement   A one- or two-page explanation of why proposed research is important or relevant.

static content   Information within a Web site that looks the same to every visitor on every visit. Compare to **dynamic content**.

stet   An editor's comment that means "ignore my comments." An editor who writes some corrections in the margin of a book and then changes his mind would write *stet* next to his corrections.

structured documentation   A guide in which the document consists of a set of chunks.

time-series graphs   A two-dimensional graph in which the *x*-axis represents the passage of time.

title The name of a document. In addition, a title is also a brief description of a table, typically prefaced by a number (for example, Table 3-1 Characteristics of Common Shade Trees). By custom, a table title appears just above each table. Compare to **caption**.

TOC Abbreviation for *table of contents*.

tone The emotional character of a document. Most technical documents have a serious, business like tone, although science writing for lay audiences often has a lighter, friendlier tone. Some modern consumer-product manuals even present a slightly flip or sarcastic tone.

transition A word or phrase that helps a reader move smoothly to the next sentence. Popular transitions in technical writing include *however, for example*, and *therefore*. When used, transitions are usually the first word or phrase in a sentence. Compare to **conjunction**.

tutorial A manual aimed at getting a neophyte started on a new topic.

variable-width font A font in which different characters occupy different amounts of horizontal space. Compare to **fixed-width font**.

Web page The information displayed at a specific URL. A collection of Web pages constitutes a Web site.

Web site An organized collection of Web pages, typically focused on providing information about a discrete topic or to serve a distinct audience.

white space A section of a hard copy document that does not contain any ink, or a section of a Web page that is rendered in the background color and pattern of that Web page.

BIBLIOGRAPHY

- Alred, G., C. Brusaw, and W. Oliu. 2000. *Handbook of Technical Writing*: 6th ed. Boston, MA: Bedford/St. Martin's.

- Chicago Editorial Staff. 2003. *Chicago Manual of Style:* 15th ed. Chicago: University of Chicago Press.

- Friedlan, A. and C. Folt. 2000. *Writing Successful Science Proposals*. New York City: Yale University Press.

- Harlow, William M. 1957. *Trees of the Eastern and Central United States and Canada.* Toronto: General Publishing Company. This book served as source for the sample tables in Chapter 8.

- Kernighan, B. and D. Richie. 1988. *The C Programming Language*: 2nd ed. Upper Saddle River, NJ: Prentice-Hall PTR.

- National Science Foundation. 2004. *Grant Proposal Guide.* Arlington, VA: July.

- Nielsen, J. 2000. *Designing Web Usability.* Indianapolis: New Riders Publishing.

- Nielsen, J. and M. Tahir. 2002. *Homepage Usability.* Indianapolis: New Riders Publishing.

- Perelman, L, J. Paradis, and E. Barrett. 1998. *The Mayfield Handbook of Technical & Scientific Writing*. Mountain View, CA: Mayfield Publishing Company.

- Rich, S. and D. Gumpert. 1985. *Business Plans That Win $$$*. New York City: Harper and Row.

- Strunk, W., E. B. White, and R. Angell. 2000. *The Elements of Style*: 4th ed. Needham, MA: Pearson.

- Tarutz, Judith. 1992. *Technical Editing.* Reading, MA: Addison-Wesley.

- Tufte, Edward. 2001. *The Visual Display of Quantitative Information*. Cheshire, CT: Graphics Press.

- Tufte, Edward. 1997. *Visual Explanations*. Cheshire, CT: Graphics Press.

Web Sites

- Purdue University. Department of Horticulture and Landscape Architecture (www.hort.purdue.edu/hort).

- Riofrio, Marianne. Ohio State University Extension Fact Sheet: Fertilizing Vegetable Garden Soils (ohioline.ag.ohio-state.edu).

smiley faces, 254
soccer in examples, 15
soft-copy documents
    editing, 265
    fonts, 276, 278
software documentation
    quotation marks, 290
    release notes, 148
software manuals
    cookbook example, 135
    example of a guide, 139
    guide example, 139
    nonverbal treatment, 142
    tutorial example, 137
sorting rows in a table, 78
speaking, 243
    fear of, 246
specifications (*see* specs)
specs, 201–213
    design, 210
    functional, 206
    high-level technical, 206
    low-level technical, 210
speech, 243
    lessons from professional entertainers,
        244
    overcoming fear, 246
    telling jokes, 230
    transitions, 57
spell checkers, 263, 264, 266
    context problems, 269
split infinitives, 13
Spring Into Series, 138
stage freight, 241
Standard General Markup Language, 280
stapler and ruler example, 113
starting a project, 128
static content in a Web site, 163
    definition of, 297
stet, 297
straight quotes, 290
strong verbs, 34
structured documentation, 138
    relation to reference manuals, 140
subheads within Web pages, 175
subjects
    active vs. passive, 46
    missing, 47
    obscuring, 48

sublists, 65
sugar, spoonful of, 261
summaries
    business proposals, 213
    doc specs, 28
    documentation project plans, 28
    editing, 269
    e-mail messages, 256
    fonts, 282
    graphics, 109
    high-level technical specs, 213
    home pages, 180
    lab reports, 226
    lists, 72
    low-level technical specs, 213
    manuals, 159
    paragraphs, 61
    PowerPoint presentations, 247
    professional secrets, 129
    proposals, 200
    secondary pages, 180
    sections, 61
    sentences, 53
    tables, 84
    Web sites, 180
    words, 43
summarizing in technical writing, 6
Sun Microsystems, 227
support organizations and release notes, 147
Swedish language, 13
synopsis
    business proposals, 204
    high-level technical specs, 207, 208
    low-level technical spec, 211

T

tables, 73–84
    amount of text in cells, 80
    audience reaction, 73
    captions, 83
    colons, 288
    column headers, 74
    editing, 263
    font height, 277
    fonts, 275
    for variety, 122
    introducing, 76

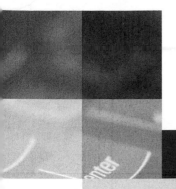

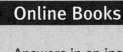